<h1 align="center">Mock Test - 1</h1>

Consider a solid sphere of density $\rho$ and radius 4R. Centre of the sphere is at origin. Two spherical cavities centered at (2R, 0) and (−2R, 0) are created in sphere. Radii of both cavities is R. In left cavity material of density $2\rho$ is filled while second cavity is kept empty. What is gravitational field at origin.

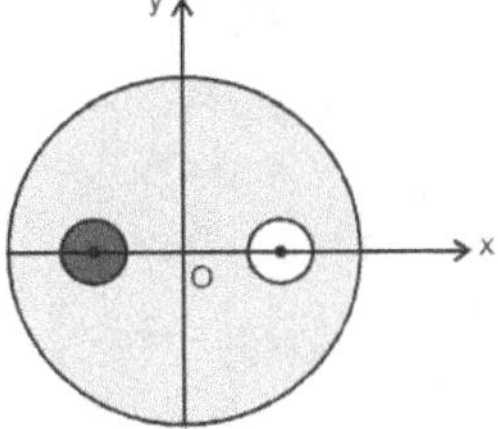

(A) $\dfrac{G\rho\pi R}{3}$

(B) $\dfrac{2G\rho\pi R}{3}$

(C) $\dfrac{4G\rho\pi R}{3}$

(D) $\dfrac{3G\rho\pi R}{2}$

The lens shown is equiconvex having refractive Index. 1.5. In the situation shown the final image of object coincides with the object. The region between lens and mirror is now filled with a liquid of Rrefractive Index 2. Then find the separation between O & image formed by convex mirror.

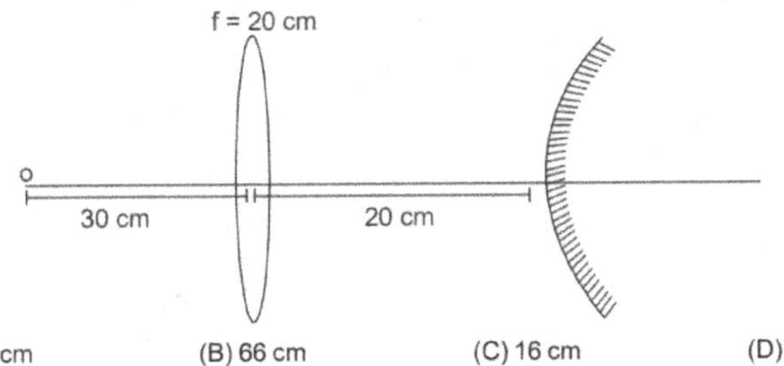


(A) 33 cm

(B) 66 cm

(C) 16 cm

(D) 32 cm

The electric field at the centre of a uniformly charged hemispherical shell is $E_0$. Now two portions of the hemisphere are cut from either side and remaining portion is shown in figure. If $\alpha = \beta = \dfrac{\pi}{3}$, then electric field intensity at centre due to remaining portion is

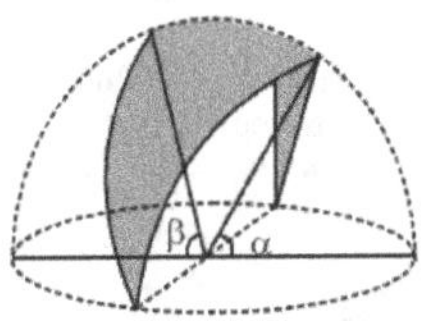


(A) $\dfrac{E_0}{3}$

(B) $\dfrac{E_0}{6}$

(C) $\dfrac{E_0}{2}$

(D) $E_0$

A thin converging lens $L_1$ forms a real image of an object located far away from the lens as shown in the figure. The image is located a distance $4\ell$ and has height h. A diverging lens of focal length $\ell$ is placed $2\ell$ from lens $L_1$. Another converging lens of focal length $2\ell$ is placed $3\ell$ from lens $L_1$. The height of final image thus formed is (Both diverging and converging lenses are placed at right side of $L_1$ -

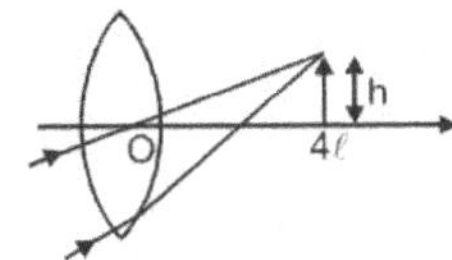


(A) h

(B) $\dfrac{h}{2}$

(C) 4h

(D) 2h

5. A point object P is moving towards left with speed 5 mm/sec parallel to optical axis of a concave mirror of focal length f = 20 cm. The seperation between object and optical axis is 1 cm. Find velocity of image of object in vector form when foot of perpendicular from object on the optical axis is at a distance 30 cm from pole.

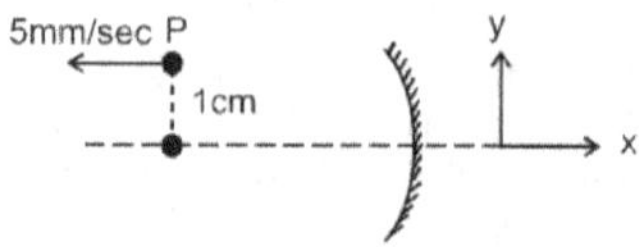


(A) $\vec{V}_i = 20\hat{i} + 4\hat{j}$ mm/sec

(B) $\vec{V}_i = 20\hat{i} + \hat{j}$ mm/sec

(C) $\vec{V}_i = 20\hat{i} - \hat{j}$ mm/sec

(D) $\vec{V}_i = 20\hat{i}$ mm/sec

6. Monochromatic light rays parallel to x-axis strike a convex lens AB. If the lens oscillates such that AB tilts upto a small angle $\theta$ (in radian) on either side of y-axis, then find the distance between extreme positions of oscillating image (f = focal length of the lens) :

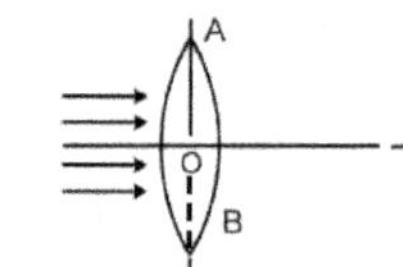


(A) $2f(\sec\theta - 1)$

(B) $f\sec^2\theta$

(C) $f(\sec\theta - 1)$

(D) the image will not move

7. Two point charges having charge +Q, –q and mass M, m respectively are separated by a distance L. They are released from rest in a uniform electric field E. The electric field is parallel to line joining both the charges and is directed from negative to positive charge. For the separation between particles to remain constant, the value of L is ($K = \dfrac{1}{4\pi\epsilon_0}$)

(A) $\sqrt{\dfrac{(M+m)KQq}{E(qM+Qm)}}$

(B) $\sqrt{\dfrac{(M+m)KQq}{E(qm+QM)}}$

(C) $\sqrt{\dfrac{mMKQq}{E(qM+Qm)}}$

(D) $\sqrt{\dfrac{mMKQq}{E(QM+qm)}}$

8. At distance 'r' from a point charge, the ratio $\dfrac{U}{V^2}$ (where 'U' is energy density and 'V' is potential) is best represented by :

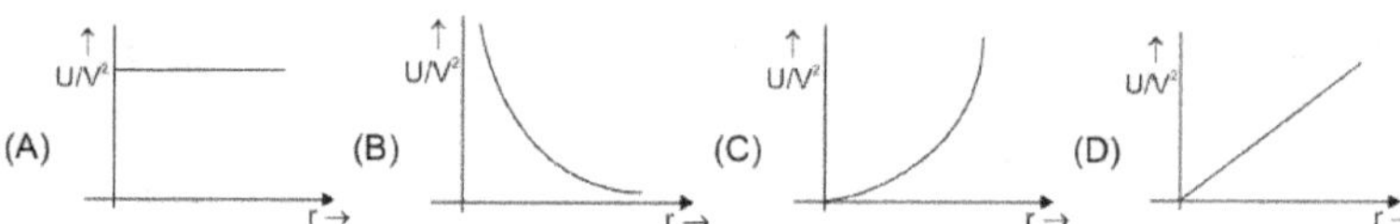


9. A cylindrical portion of radius r is removed from a solid sphere of radius R and uniform volume charge density $\rho$ in such a way that the axis of the hollow cylinder coincides with one of the diameters of the sphere. (r is negligible compared to R). Then the electric field intensity at point A is

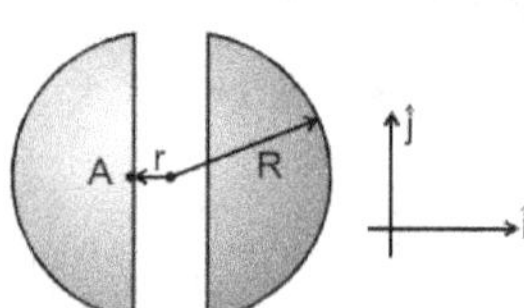


(A) $\dfrac{\rho r}{3\varepsilon_0}\hat{i}$

(B) $-\dfrac{\rho r}{3\varepsilon_0}\hat{i}$

(C) $\dfrac{\rho r}{6\varepsilon_0}\hat{i}$

(D) $-\dfrac{\rho r}{6\varepsilon_0}\hat{i}$

Two satellites revolve around the 'Sun' as shown in the figure. First satellite revolves in a circular orbit of radius R with speed $v_1$. Second satellite revolves in elliptical orbit, for which minimum and maximum distance from the sun are $\dfrac{R}{3}$ and $\dfrac{5R}{3}$ respectively. Velocities at these positions are $v_2$ and $v_3$ respectively. The correct order of speeds is

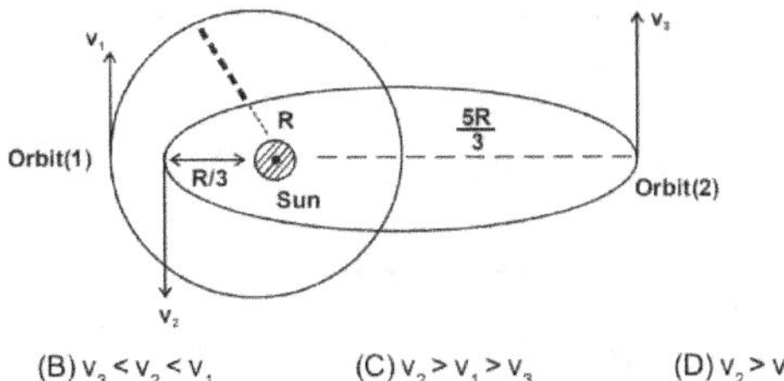


(A) $v_2 > v_3 > v_1$      (B) $v_3 < v_2 < v_1$      (C) $v_2 > v_1 > v_3$      (D) $v_2 > v_3 = v_1$

A small area is removed from a uniform spherical shell of mass M and radius R. Then the gravitational field intensity near the hollow portion is

(A) $\dfrac{GM}{R^2}$      (B) $\dfrac{GM}{2R^2}$      (C) $\dfrac{3GM}{2R^2}$      (D) Zero

A meteorite approaching a planet of mass M (in the straight line passing through the centre of the planet) collides with an automatic space station orbiting the planet in a circular trajectory of radius R. The mass of the station is ten times as large as the mass of the meteorite. As a result of the collision, the meteorite sticks in the station which goes over to a new orbit with the minimum distance R/2 from the planet. Speed of the meteorite just before it collides with the planet is : .

(A) $\sqrt{\dfrac{58GM}{R}}$      (B) $\sqrt{\dfrac{38GM}{R}}$      (C) $\sqrt{\dfrac{28GM}{R}}$      (D) $\sqrt{\dfrac{18GM}{R}}$

Two converging lenses have focal length $f_1$ and $f_2$ $(f_1 > f_2)$. The optical axis of the two lenses coincide. This lens system is used to from an image of real object. It is observed that final magnification of the image does not depend on the distance x. Whole arrangement is shown in figure. Final magnification is :

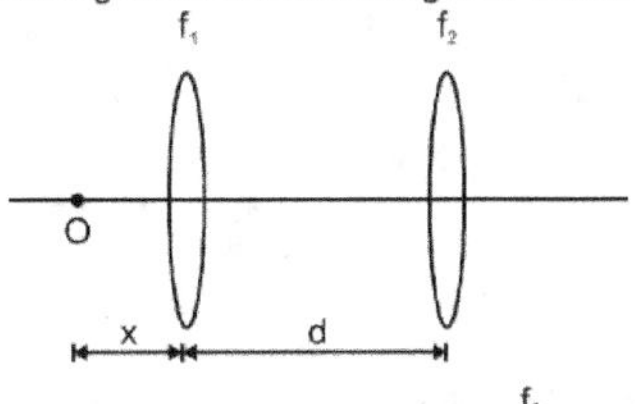


(A) $-\dfrac{f_1}{f_2}$      (B) $-\dfrac{f_2}{f_1}$      (C) $-\dfrac{f_1}{f_1+f_2}$      (D) $-\dfrac{f_2}{f_1-f_2}$

In the figure shown an infinitely long wire of uniform linear charge density $\lambda$ is kept perpendicular to the plane of figure such that it extends upto infinity on both sides of the paper. Find the electrostatic force on a semicircular ring kept such that its geometrical axis coincides with the wire. The semicircular ring has a uniform linear charge density $\lambda'$.

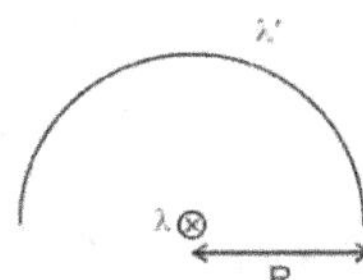


(A) $\dfrac{\lambda\lambda'}{\pi\varepsilon_0 R}$      (B) $\dfrac{\lambda\lambda'}{4\pi\varepsilon_0}$      (C) $\dfrac{\lambda\lambda'}{2\pi\varepsilon_0}$      (D) $\dfrac{\lambda\lambda'}{\pi\varepsilon_0}$

15. Consider a spherical planet rotating about its axis. The velocity of a point at equator is v. The angular velocity of this planet is such that it makes apparent value of 'g' at the equator half of value of 'g' at the pole. The escape speed for a polar particle on the planet expressed as multiple of v is :

(A) v          (B) 2v          (C) 3v          (D) 4v

16. The figure shows two equal, positive charges, each of magnitude 50 $\mu$C, fixed at points (3, 0) m and (−3, 0)m respectively. A charge −50$\mu$C, moving along negative y–axis has a kinetic energy of 4J at the instant it crosses point (0,4)m. Determine the position of this charge where the direction of its motion reverses for the first time after crossing this point (neglect gravity).

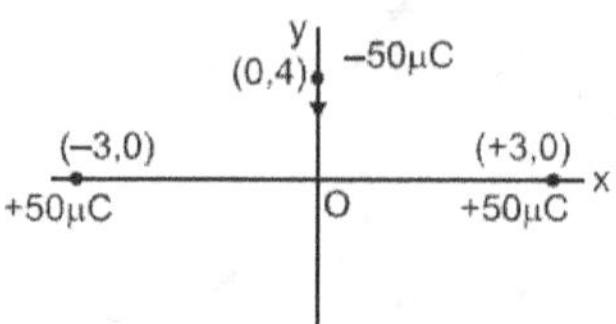


(A) $\left(0m, -7\sqrt{2}m\right)$      (B) $\left(0, -6\sqrt{2}m\right)$      (C) $(0, -5\sqrt{2})$      (D) $\left(0m, -4\sqrt{2}m\right)$

17. Orbital velocity of a satellite in its orbit (around earth) of radius r is v. It collides with another body in its orbit and comes to rest just after the collision. Taking the radius of earth as R, the speed with which it will fall on the surface of earth will be :

(A) $v\sqrt{(\frac{r}{R}-1)}$      (B) $v\sqrt{2(\frac{r}{R}-1)}$      (C) $\dfrac{v}{\sqrt{2(\frac{r}{R}+1)}}$      (D) $v\sqrt{2(\frac{r}{R}+1)}$

18. A solid spherical planet of mass 2m and radius 'R' has a very small tunnel along its diameter. A small cosmic particle of mass m is at a distance 2R from the centre of the planet as shown. Both are initially at rest, and due to gravitational attraction, both start moving toward each other. After some time, the cosmic particle passes through the centre of the planet. (Assume the planet and the cosmic particle are isolated from other planets)

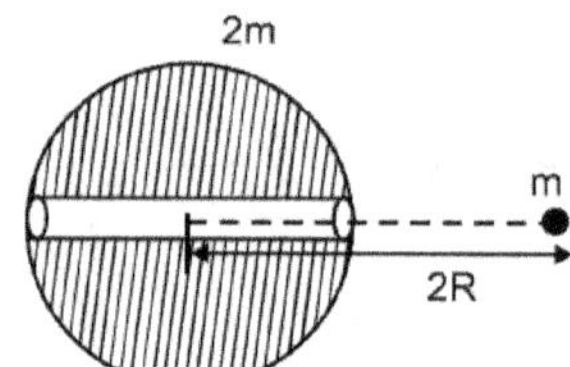


(A) Displacement of the cosmic particle till that instant is $\dfrac{4R}{3}$

(B) Acceleration of the cosmic particle at that instant is zero

(C) velocity of the cosmic particle at that instant is $\sqrt{\dfrac{8Gm}{3R}}$

(D) Total work done by the gravitational force on both the particle is $-\dfrac{2Gm^2}{R}$

In the figure shown A & B are two charged particles having charges q and – q respectively are placed on a non-conducting fixed horizontal smooth plane. B is fixed and A is attached to a non conducting massless spring of spring constant k. The other end of the spring is fixed. Mass of A is m, A and B are in equilibrium when the distance between them is r. Choose the correct options

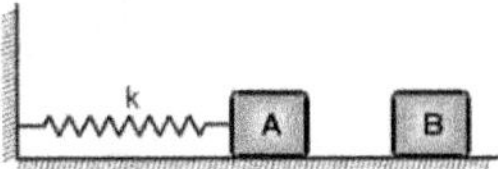


(A) time period of small oscillation of block A about is means position $= 2\pi\sqrt{\dfrac{m}{k - \dfrac{q^2}{2\pi\epsilon_0 r^3}}}$.

(B) time period of small oscillation of block A about is means position $= 2\pi\sqrt{\dfrac{m}{k + \dfrac{q^2}{2\pi\epsilon_0 r^3}}}$.

(C) to perform SHM, K must be greater than $\left(\dfrac{q^2}{2\pi\epsilon_0 r^3}\right)$.

(D) to perform SHM, K must be greater than $\left(\dfrac{2q^2}{\pi\epsilon_0 r^3}\right)$

A light ray enters into a medium whose refractive index varies along the x-axis as $n(x) = n_0\sqrt{1 + \dfrac{x}{4}}$ where $n_0 = 1$. The medium is bounded by the planes $x = 0$, $x = 1$ & $y = 0$. If the ray enters at the origin at an angle $30°$ with x–axis.

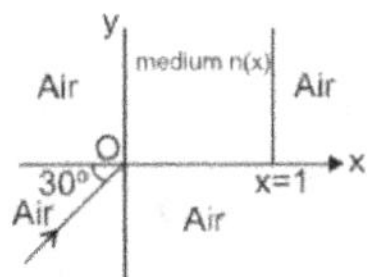


(A) equation of trajectory of the light ray is $y = [\sqrt{3+x} - \sqrt{3}]$

(B) equation of trajectory of the light ray is $y = 2[\sqrt{3+x} - \sqrt{3}]$

(C) the coordinate the point at which light ray comes out from the medium is $[1, 2(2-\sqrt{3})]$

(D) the coordinate the point at which light ray comes out from the medium is $[0, 2(2-\sqrt{3})]$

A charge 'q' is placed on the diagonal AP of a cube at a distance $\dfrac{AP}{3}$ from the point A. Choose the correct options.

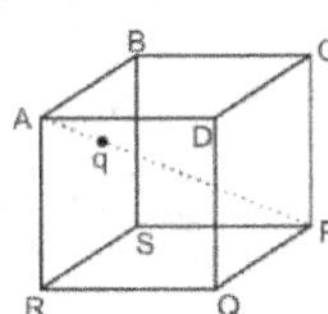


(A) the sum of electric flux passing through the surfaces ABCD and PQRS is $\dfrac{q}{3\varepsilon_0}$

(B) the sum of electric flux passing through the surfaces ABCD and PQRS is $\dfrac{q}{8\varepsilon_0}$

(C) the flux through both the surfaces ABCD and PQRS are same

(D) the flux through the surfaces ABCD is larger than the flux through surface PQRS.

Two infinite, parallel, non–conducting thin sheets carry equal positive charge density $\sigma$. One is placed in the yz plane and the other at $x = a$. Take potential $V = 0$ at $x = 0$. Choose the correct statements

(A) For $0 < x < a$, potential $V = 0$.

(B) For $x > a$, potential $V = -\dfrac{\sigma}{\epsilon_0}(x-a)$

(C) For $x > a$, potential $V = \dfrac{\sigma}{\epsilon_0}(x-a)$

(D) For $x < 0$ potential $V = \dfrac{\sigma}{\epsilon_0}x$

23. In the figure shown there is a hollow hemisphere of radius 'R'. It has a uniform mass distribution having total mass m. The gravitational potential at points A, D and B are $V_A$, $V_D$ and $V_B$ respectively. Distance of D and B from centre C are R/2 and 2R respectively. The points C, D and B are lying on radial line of the hollow hemisphere.

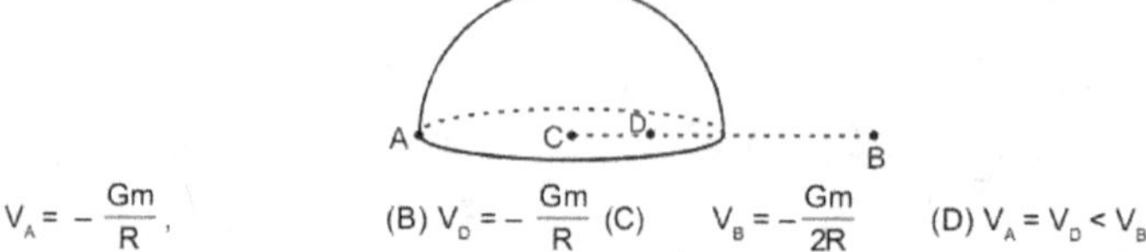


(A) $V_A = -\dfrac{Gm}{R}$ ,    (B) $V_D = -\dfrac{Gm}{R}$   (C)   $V_B = -\dfrac{Gm}{2R}$   (D) $V_A = V_D < V_B$

24. A cavity of radius r is present inside a fixed solid dielectric sphere of radius R, having a volume charge density of $\rho$. The distance between the centres of the sphere and the cavity is a. An electron is released inside the cavity at an angle $\theta = 45^0$ as shown. The electron (of mass m and charge –e) will take $\left(\dfrac{P\sqrt{2}\, m\, r\, \varepsilon_0}{e\, a\, \rho}\right)^{1/2}$ time to touch the sphere again. Neglect gravity. Find the value of P :

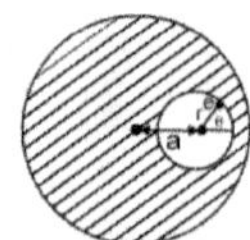

25. A planet revolves around the sun in elliptical orbit of semimajor axis $2 \times 10^{12}$ m. The areal velocity of the planet when it is nearest to the sun is $4.4 \times 10^{16}$ m²/s. The least distance between planet and the sun is $1.8 \times 10^{12}$ m. Find the minimum speed of the planet in km/s.

26. A light ray parallel to the principal axis is incident (as shown in the figure) on a planoconvex lens with radius of curvature of its curved part equal to 10 cm. Assuming that the refractive index of the material of the lens is 4/3 and medium on both sides of the lens is air, the distance of the point from the lens where this ray meets the principal axis is $\dfrac{2y}{7}$ cm then find value of y.

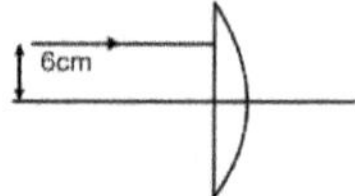


27. A satellite is orbiting around the earth in a circular orbit and in this orbit magnitude of its acceleration is '$a_1$'. Now a rocket is fired in the direction of motion of satellite from the satellite due to which its speed instantaneously becomes half of initial, just after the rocket is fired acceleration of satellite has magnitude '$a_2$'. Then the ratio $\dfrac{a_1}{a_2}$ is (Assume there is no external force other than the gravitational force of earth before and after the firing of rocket from the satellite)

28. A uniform thin rod of mass m and length R is placed normally on surface of earth as shown. The mass of earth is M and its radius is R. If the magnitude of gravitational force exerted by earth on the rod is $\dfrac{\eta GMm}{12R^2}$ , then '$\eta$' is

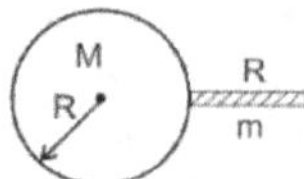


29. Two persons A and B wear glasses of optical powers (in air) $P_1 = + 2$ D and $P_2 = + 1$ D respectively. The glasses have refractive index 1.5. Now they jump into a swimming pool and look at each other. B appears to be present at distance 2m (from A) to A. A appears to be present at distance 1m (from B) to B. The refractive index of water in the swimming pool, in the form $\dfrac{X}{10}$ and find X.

The final image I of the object O shown in the figure is formed at point 20 cm below a thin equi-concave lens, which is at a depth of 65 cm from principal axis. From the given geometry, calculate the radius of curvature in cm of lens kept at "A". (Refractive index of equi-convex lens is 1.5 and placed in air.

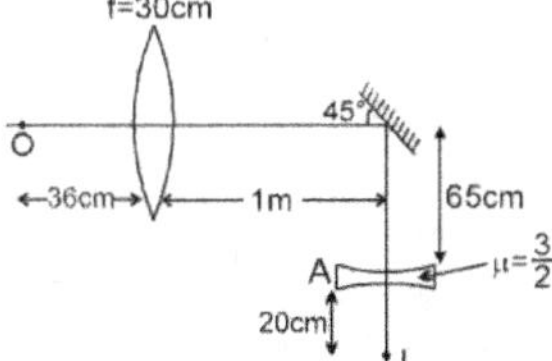


A planet is made of two materials of density $\rho_1$ and $\rho_2$ as shown in figure.

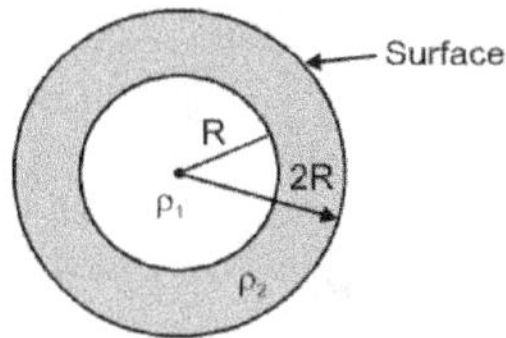


Acceleration due to gravity at surface of planet is same as at depth 'R'. The ratio $\dfrac{\rho_1}{\rho_2}$ is equal to $\dfrac{x}{3}$. Find the value of x.

Figure shows an irregular block of material of refractive index $\sqrt{2}$. A ray of light strikes the face AB as shown in figure. After refraction it is incident on a spherical surface CD of radius of curvature 0.4 m, (with centre lying on the line PQ) and enter a medium of refractive index 1.514 to meet PQ at E. Find the distance OE. in nearest interger in meters (point M is very near to line PQ)

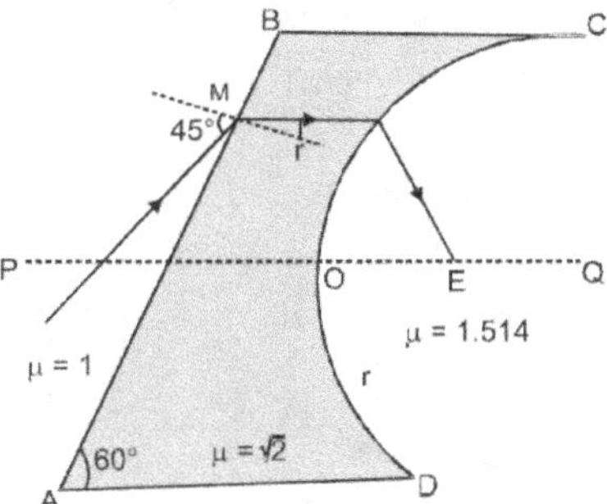

# COMPREHENSION -1

The curve of angle of incidence versus angle of deviation shown has been plotted for prism.

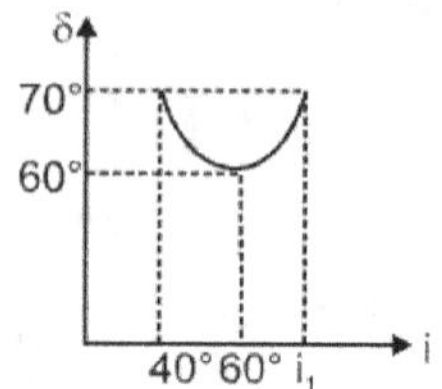

33. The value of refractive index of the prism used is

(A) $\sqrt{3}$       (B) $\sqrt{2}$       (C) $\sqrt{3}/\sqrt{2}$       (D) $2/\sqrt{3}$

34. The value of angle $i_1$ in degrees is
(A) $40°$       (B) $60°$       (C) $70°$       (D) $90°$

## ΟMPREHENSION -2

There are two non-conducting spheres having uniform volume charge densities $\rho$ and $-\rho$. Both spheres have equal radius R. The spheres are now laid down such that they overlap as shown in the figure.

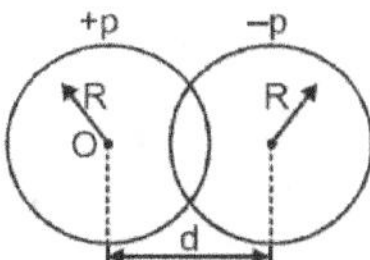


The electric field E in the overlap region is

(A) non uniform      (B) zero      (C) $\dfrac{\rho}{3\,\epsilon_0}d$      (D) $\dfrac{\rho}{3\,\epsilon_0}R$

The potential difference $\Delta V$ between the centers of the two spheres for d = R is :

(A) $\dfrac{\rho}{3\,\epsilon_0}d^2$      (B) $\dfrac{\rho}{\epsilon_0}d^2$      (C) zero      (D) $\dfrac{2\rho}{\epsilon_0}d^2$

## ΟMPREHENSION -3

Consider a hypothetical solar system, which has two identical massive suns each of mass M and radius r,

seperated by a seperation of $2\sqrt{3}\,R$ (centre to centre). (R >>>r). These suns are always at rest. There is only one planet in this solar system having mass m. This planet is revolving in circular orbit of radius R such that centre of the orbit lies at the mid point of the line joining the centres of the sun and plane of the orbit is perpendicular to the line joining the centres of the sun. Whole situation is shown in the figure.

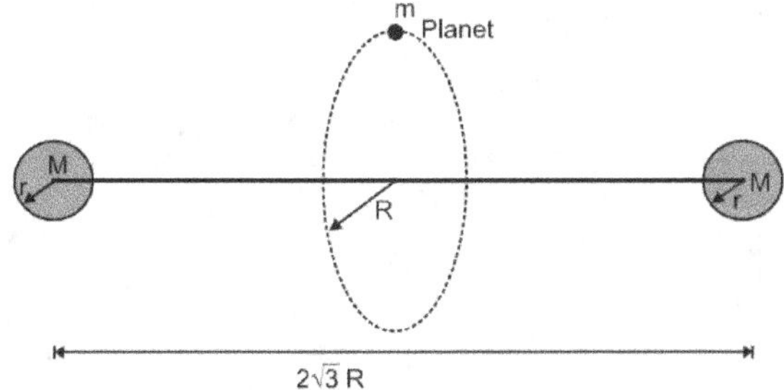


## Answer the following qustion regarding to this solar system.

7.     Speed of the planet is :

(A) $\sqrt{\dfrac{GM}{8R}}$          (B) $\sqrt{\dfrac{GM}{4R}}$

(C) $\sqrt{\dfrac{GM}{2R}}$          (D) $\sqrt{\dfrac{GM}{3R}}$

38. **Average force on the planet in half revolution is :**

(A) $\dfrac{GMm}{4\pi R^2}$

(B) $\dfrac{GMm}{4R^2}$

(C) $\dfrac{GMm}{2\pi R^2}$

(D) $\dfrac{GMm}{8R^2}$

39. **Duration of one year for this planet is :**

(A) $\dfrac{4\pi R^{3/2}}{\sqrt{GM}}$

(B) $\dfrac{2\pi R^{3/2}}{\sqrt{GM}}$

(C) $\dfrac{\pi R^{3/2}}{\sqrt{GM}}$

(D) $\dfrac{3\pi R^{3/2}}{\sqrt{GM}}$

**COMPREHENSION -4**

A charge q is divided into three equal parts and placed symmetrically on a circle of radius r. The same charge is divided into four equal parts and placed symmetrically on the same circle. The electric field intensities at the centre of the circle in two situations are zero.

40. The ratio of electric potentials at the centre in the two situations is

(A) $\dfrac{2}{\sqrt{3}}$
(B) $\dfrac{1}{1}$
(C) $\dfrac{4}{3}$
(D) $\dfrac{16}{9}$

41. The potential energy of the system in first situation where the charge is divided into three equal parts is

(A) $\dfrac{1}{4\pi\varepsilon_0}\dfrac{q^2}{r}$
(B) $\dfrac{1}{36\pi\varepsilon_0}\dfrac{q^2}{r}$
(C) $\dfrac{1}{12\sqrt{3}\pi\varepsilon_0}\dfrac{q^2}{r}$
(D) $\dfrac{1}{12\pi\varepsilon_0}\dfrac{q^2}{r}$

42. If a charge (part charge) is removed from one location in both the situations, the ratio of magnitudes of the electric field intensities at the centre is

(A) $\dfrac{1}{2}$
(B) $\dfrac{1}{1}$
(C) $\dfrac{2}{3}$
(D) $\dfrac{4}{3}$

Match the proper entries from column-2 to column-1 using the codes given below the columns, if deviation in the Column–II is the magnitude of total deviation (between incident ray and finally refracted or reflected ray) to lie between 0° and 180°. Here n represents refractive index of medium.

| Column–I | Column–II |
|---|---|

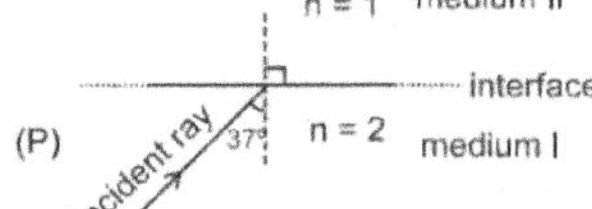


(1) deviation in the light ray is greater than 90°

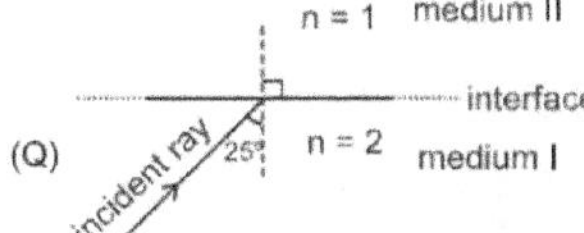


(2) deviation in the light ray is less than 90°

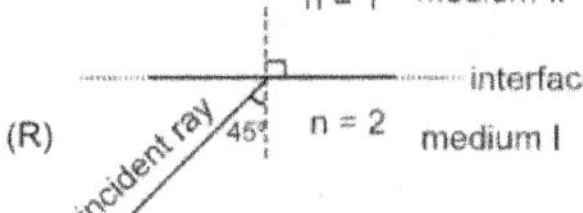


(3) deviation in the light ray is equal to 90°

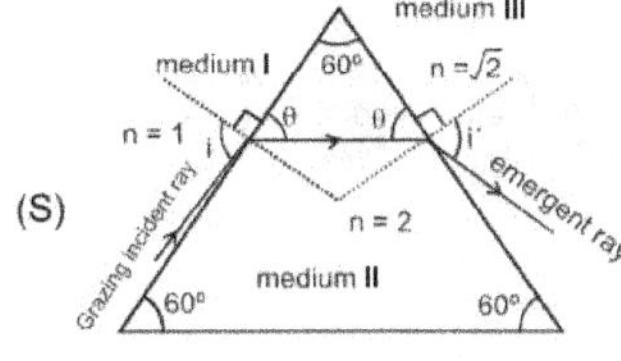


(4) Speed of finally reflected or refracted light is

same as speed of incident light.

|  | (P) | (Q) | (R) | (S) |
|---|---|---|---|---|
| (A) | 4 | 2 | 1 | 3 |
| (B) | 1 | 2 | 4 | 2 |
| (C) | 3 | 1 | 4 | 2 |
| (D) | 2 | 4 | 1 | 3 |

**44.** In each situation of column-I, some charge distributions are given with all details explained. In column -II The electrostatic potential energy and its nature is given situation in column -II. Match the proper entries from column-2 to column-1 using the codes given below the columns,

**Column-I**

(P) A thin shell of radius a and having a charge $-Q$ uniformly distributed over its surface as shown

(Q) A thin shell of radius $\dfrac{5a}{2}$ and having a charge $-Q$ uniformly distributed over its surface and a point charge $-Q$ placed at its centre as shown.

(R) A solid sphere of radius a and having a charge $-Q$ uniformly distributed throughout its volume as shown.

(S) A solid sphere of radius a and having a charge $-Q$ uniformly distributed throughout its volume. The solid sphere is surrounded by a concentric thin uniformly charged spherical shell of radius 2a and carrying charge $-Q$ as shown

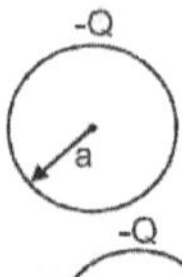


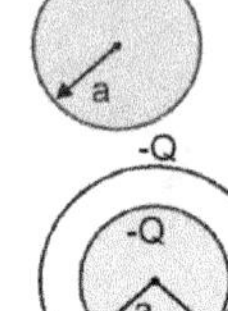


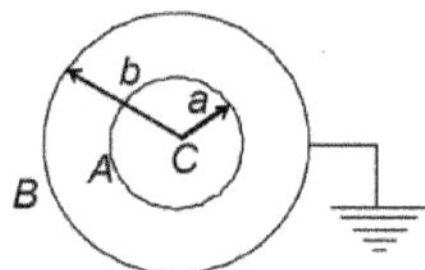




**Column-II**

(1) $\dfrac{1}{8\pi\,\epsilon_0}\dfrac{Q^2}{a}$ in magnitude

(2) $\dfrac{3}{20\pi\,\epsilon_0}\dfrac{Q^2}{a}$ in magnitude

(3) $\dfrac{27Q}{80\pi\,\epsilon_0}\dfrac{Q^2}{a}$ in magnitude

(4) Positive in sign

|     | (P) | (Q) | (R) | (S) |
|-----|-----|-----|-----|-----|
| (A) | 1   | 2   | 4   | 3   |
| (B) | 4   | 1   | 2   | 3   |
| (C) | 1   | 2   | 3   | 4   |
| (D) | 2   | 4   | 3   | 1   |

**45.** A conducting sphere $A$ of radius a, with charge $Q$ is placed concentrically inside a conducting shell $B$ of radius $b$. $B$ is earthed, $C$ is the common centre of $A$ and $B$. If $P$ is the point between shells $A$ and $B$ at distance $r$ from center $C$ then Match the proper entries from column-2 to column-1 using the codes given below the columns,

(use : $a = 1m$, $b = 3m$, $r = 2$ m and $K = \dfrac{1}{4\pi\varepsilon_0}$ )

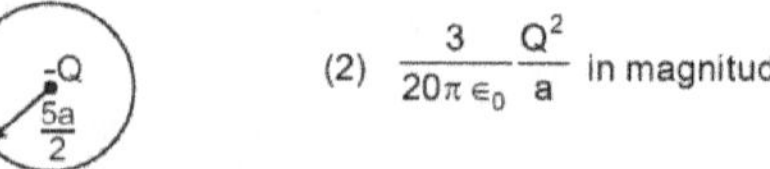


**Column - I**

(P) Electric field at point $P$ is

(Q) Electric potential at point $P$ is $(v_\infty = 0)$

(R) Electric potential difference between $A$ and $B$ is

(S) Electric field outside the shell $B$ at distance 5 m from centre $C$ is

**Column - II**

(1) $\dfrac{2KQ}{3}$

(2) zero

(3) $\dfrac{KQ}{4}$

(4) $\dfrac{KQ}{6}$

|     | (P) | (Q) | (R) | (S) |
|-----|-----|-----|-----|-----|
| (A) | 3   | 1   | 2   | 4   |
| (B) | 1   | 2   | 3   | 4   |
| (C) | 2   | 3   | 4   | 1   |
| (D) | 3   | 4   | 1   | 2   |

# olutions

Above distribution can be represented as shown in figure.
Gravitational field due to sphere of radius R at a distance 2R

$$E_g = \frac{G\rho \frac{4}{3}\pi R^3}{4R^2} = \frac{G\rho\pi R}{3}$$

So Net field at centre will be $2F_g = \dfrac{2G\rho\pi R}{3}$

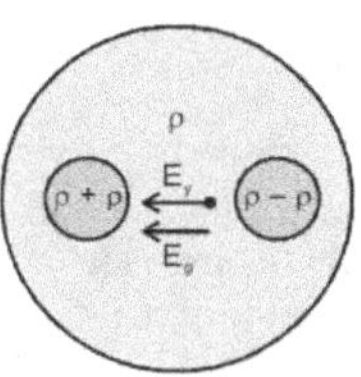


**Case-I**
Radius of curvature of lens is 20 cm
Image formed by convex lens should be at centre of curvature of mirror

$$\frac{1}{V} + \frac{1}{30} = \frac{1}{20}$$

$$\frac{1}{V} = \frac{1}{20} - \frac{1}{30} \Rightarrow V = 60 \text{ cm}$$

Radius curvature of mirror should be 40 cm.
**Case-II**

$$\frac{2}{V_1} + \frac{1}{30} = \frac{1.5-1}{20} + \frac{2-1.5}{-20}$$

$$\Rightarrow V = -60$$

So for convex mirror $u = -80$

$$\frac{1}{V} - \frac{1}{80} = \frac{1}{20}$$

$V = 16$ cm
Seperation between object and this image O = 66 cm

Consider the whole hemisphere as three portion if electric field due to one portion is $E_1$
then $2E_1 \sin 30 + E_1 = E_0$

$$2E_1 = E_0$$

$$\Rightarrow E_1 = \frac{E_0}{2}$$

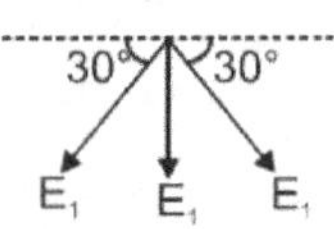


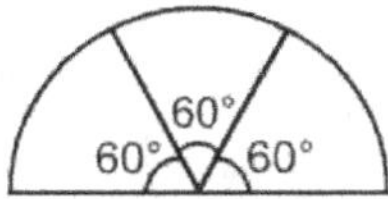


From 2$^{nd}$ lens $\qquad \dfrac{1}{V} - \dfrac{1}{2\ell} = \dfrac{1}{-\ell}$ or $\qquad v = -2\ell$

$m_1 = -1$

From 3$^{rd}$ lens $\qquad \dfrac{1}{V} - \dfrac{1}{-3\ell} = \dfrac{1}{2\ell}$ or $\qquad v = 6\ell$

$m_2 = -2$
$h_i = (m_1 \times m_2) h_0$
$= 2h$

**5.**

$$\frac{1}{v} + \frac{1}{-30} = \frac{1}{-20}$$

$$v = -60$$

$$m = \frac{y_i}{y_o} = \frac{v}{u}$$

$$y_i = -2 \text{ cm}$$

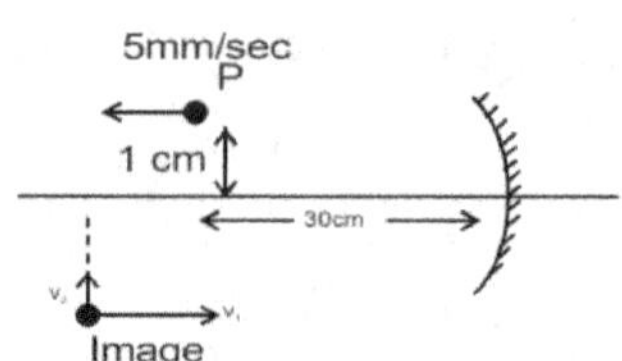


for $\vec{v}_1$ $\vec{v}_1$ $\qquad \vec{v}_1 = -\dfrac{v^2}{u^2}(\vec{v}_p)$

$$= -4\,(-5) = 20 \text{ mm/sec}$$

for $\vec{v}_2$ $\vec{v}_2$ $\qquad \Rightarrow \qquad \dfrac{y_i}{y_o} = \dfrac{v}{u}$

$$y_i\, u = -y_o\, v$$

$$\frac{dy_i}{dt}(u) + y_i\,\frac{du}{dt} = -y_o\,\frac{dv}{dt}$$

$$\frac{dy_i}{dt}(-30) + (-2)(-5) = -(20)$$

$$\frac{dy_i}{dt} = 1 \text{mm/sec}$$

$$V_i = 20\hat{i} + \hat{j} \text{ mm/sec} \qquad \textbf{Ans.}$$

**6.** When the lens is tilted by $\theta$, the image is formed at the intersection (Q) of focal plane of lens in tilted position and x-axis.

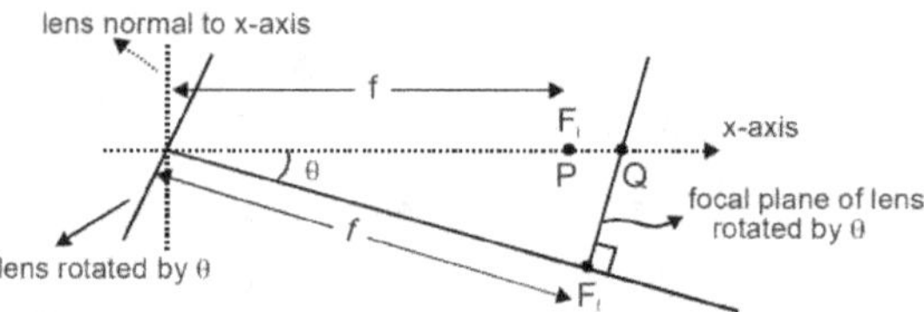


As the lens oscillates. The image shifts on x-axis in between P and Q.

$\therefore$ Distance between two extreme position of the image = PQ = $\dfrac{f}{\cos\theta} - f = f(\sec\theta - 1)$ **Ans.**

**7.** In order to maintain constant separation, the particles must have the same acceleration. Assuming the system of both charges to accelerate towards left. Applying Newton's second law.



$$QE - \frac{KQq}{L^2} = Ma \qquad \qquad \dots (1)$$

Under given condition the acceleration of both charges should be same and should also be equal to acceleration of centre of mass of both the charges.

$$a = \frac{F_{net}}{\text{total mass}} = \frac{(Q-q)E}{m+M} \qquad \dots (2)$$

Hence from equation (1) and (2) we get $\qquad L = \sqrt{\dfrac{(M+m)KQq}{E(qM+Qm)}}$

$$U = \frac{1}{2}\,\varepsilon_0 E^2 = \frac{1}{2}\,\frac{\varepsilon_0 K^2 Q^2}{r^4}$$

$$V = \frac{KQ}{r}$$

$$\frac{U}{V^2} = \frac{\dfrac{1}{2}\varepsilon_0 K^2 \dfrac{Q^2}{r^4}}{\dfrac{K^2 Q^2}{r^2}} = \frac{1}{2}\frac{\varepsilon_0}{r^2}$$

because $\dfrac{U}{V^2} \propto \dfrac{1}{r^2}$

so the correct option is B.

Field at A
due to the solid sphere without the cylindrical cavity

$$E_1 = -\frac{\rho r}{3\varepsilon_0}\,\hat{i}$$

field at A due to the cylinder of length 2R (which can be assumed to be infinite, since r << R)

$$E_2 = \frac{2K(\rho \pi r^2)}{r}(-\hat{i}) = -\frac{\rho}{2\varepsilon_0}r\,\hat{i}$$

$\therefore$ net field $E = E_1 - E_2 = \dfrac{\rho r}{6\varepsilon_0}\,\hat{i}$

$$V_1 = \sqrt{\frac{GM}{R}} \qquad \text{(orbital velocity in circular path)}$$

For elliptical orbit

conservation of angular momentam $mV_2\dfrac{R}{3} = \dfrac{5R}{3}mV_3$

conservation of energy $-\dfrac{GMm}{R/3} + \dfrac{1}{2}mV_2^2 = \dfrac{-GMm}{5R/3} + \dfrac{1}{2}mV_3^2$

Solving $V_2 = \sqrt{\dfrac{5GM}{R}}$ and $V_3 = \sqrt{\dfrac{GM}{5R}}$

**11.** Consider a small area (shaded strip)
here $E_{self}$ = Gravitational field due to this strip
and $E_{ext}$ = Gravitational field due to the rest of spherical shell.
$E_{in}$ = Gravitational field just inside the strip due to whole shell.
$E_{out}$ = Gravitational field just outside the strip due to whole shell.

$$E_{in} = E_{ext} - E_{self} = 0$$
$$\Rightarrow E_{ext} = E_{self}$$

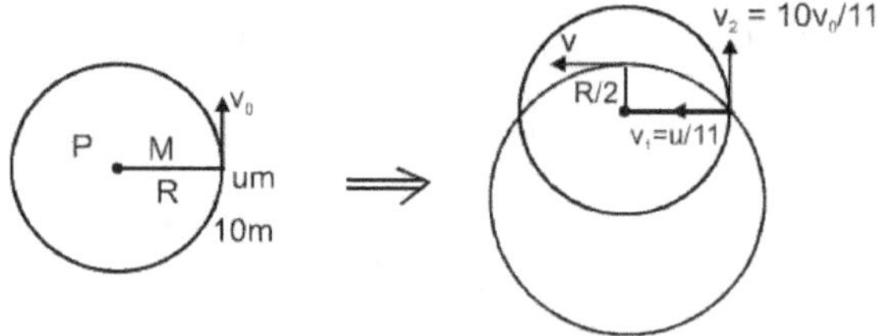


$$E_{out} = E_{ext} + E_{self} = \frac{GM}{R^2} \qquad \Rightarrow E_{ext} = \frac{GM}{2R^2}$$

After the shaded area has been removed there is no $E_{self}$ and only $E_{ext}$.

hence, $E_{net} = E_{ext} = \dfrac{GM}{2R^2}$

**12.** A s the space station is moving in circular orbit,



$$\frac{GM(10m)}{R^2} = \frac{(10m)\, v_0^2}{R}$$

$$\Rightarrow \quad v_0 = \sqrt{\frac{GM}{R}} \qquad\qquad ...(i)$$

Let u be the velocity of meteorite.
Velocity of the space station after collision can be obtained from momentum conservation.

$$mu = (10m + m)\, v_1 \qquad \Rightarrow \qquad v_1 = \frac{u}{11}$$

$$10\,m\,.\,v_0 = (10\,m + m)\, v_2 \qquad \Rightarrow \qquad v_2 = \frac{10}{11}\, v_0$$

Let v be the velocity of space station at closest distance
from angular momentum conservation

$$10\,m\,v_0 \times R = 11\,mv\,\frac{R}{2} \quad \Rightarrow \quad v = \frac{20 v_0}{11}$$

from energy conservation

$$\frac{1}{2} \times (11\,m)(v_1^2 + v_2^2) - \frac{GM\,(11\,m)}{R} = \frac{1}{2} \times (11m)\, v^2 - \frac{GM.11m}{R/2}$$

$$\Rightarrow \quad \left(\frac{u}{11}\right)^2 + \left(\frac{10 v_0}{11}\right)^2 - \frac{2GM}{R} = \left(\frac{20 v_0}{11}\right)^2 - \frac{4GM}{R}$$

$$\Rightarrow \qquad \frac{u^2}{11^2} = \frac{400\,v_0^2}{11^2} - \frac{100\,v_0^2}{11^2} - \frac{2GM}{R}$$

$$\Rightarrow \qquad u^2 = \frac{GM}{R}(400 - 100 - 242) = 58\,\frac{GM}{R}$$

Ans: $u = \sqrt{\dfrac{58\,GM}{R}}$

**13.**  Image -1

$u_1 = -x$

$$\frac{1}{v_1} - \frac{1}{-x} = \frac{1}{f_1}$$

$$v_1 = \frac{xf_1}{x-f_1}$$

$$m_1 = \frac{v_1}{u_1} = \frac{v_1}{-x} = -\left(\frac{f_1}{x-f_1}\right)$$

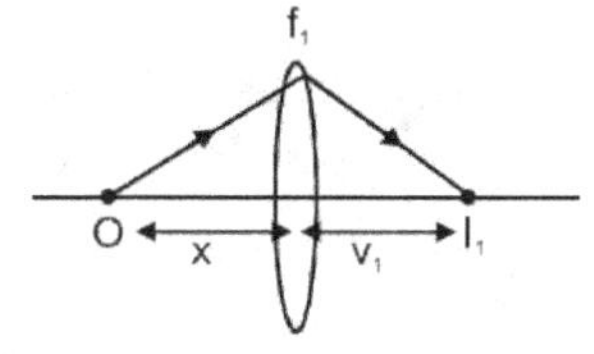


Image -2

$u_2 = -(d - v_1)$

$$\frac{1}{v_2} - \frac{1}{-(d-v_1)} = \frac{1}{f_2}$$

$$v_2 = \frac{(d-v_1)f_2}{d-v_1-f_2}$$



$$m_2 = \frac{v_2}{-(d-v_1)} = -\left(\frac{f_2}{d-v_1-f_2}\right)$$

$$m_1 m_2 = \left(\frac{f_1}{x-f_1}\right)\left(\frac{f_2}{d-\dfrac{xf_1}{(x-f_1)}-f_2}\right) = \frac{f_1 f_2}{x(d-f_1-f_2)-df_1+f_1 f_2}$$

Since m is independent of x

$$\Rightarrow \qquad (d-f_1-f_2) = 0 \Rightarrow d = f_1 + f_2$$

$$\Rightarrow \qquad m = -\frac{f_2}{f_1}$$

14. The electrostatic field intensity at a point on the ring is $E = \dfrac{\lambda}{2\pi\varepsilon_0}\,\dfrac{1}{R}$.

The force on the elementary charge dq is

$$dF = dq\,E = (\lambda'\,Rd\theta)\cdot\dfrac{\lambda}{2\pi\varepsilon_0}\,\dfrac{1}{R}$$

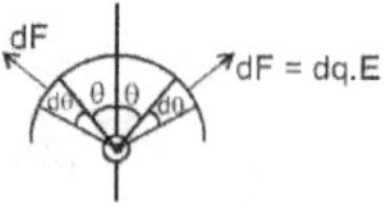


The sine component of dF will get cancelled and cosine component will get added. Net force on the ring

$$F = 2\int_0^{+\pi/2} dF\cos\theta = 2\int_0^{+\pi/2}\dfrac{\lambda.\lambda'}{2\pi\varepsilon_0}\,d\theta.\cos\theta = \dfrac{\lambda\lambda'}{\pi\varepsilon_0}$$

**Ans.** $\dfrac{\lambda\lambda'}{\pi\varepsilon_0}$

15. According to question (At equator)

$$Mg - \dfrac{Mv^2}{R} = \dfrac{Mg}{2} \qquad \Rightarrow \qquad v^2 = \dfrac{Rg}{2} = \dfrac{GM}{2R}$$

Using conservation of energy : $-\dfrac{GMm}{R} + \dfrac{1}{2}mv_e^2 = 0 \qquad \Rightarrow \qquad v_e^2 = \dfrac{2GM}{R} = 4v^2$

16. The charge $-50\mu C$ will move in straight line along y–axis as it does not experience any force in x–direction. Let B be the location where the charge comes to rest momentarily and then return. Total energy of the system remain constant.

$\therefore \quad$ KE + PE

$$= \quad 4 + \dfrac{1}{4\pi\varepsilon_0}\dfrac{(50\times10^{-6})(-50\times10^{-6})}{5}\times2$$

$$= \quad 0 + \dfrac{1}{4\pi\varepsilon_0}\dfrac{(50\times10^{-6})(-50\times10^{-6})}{\sqrt{3^2+y^2}}\times2$$

$$\dfrac{1}{4\pi\varepsilon_0} = 9\times10^9\ Nm^2\,C^{-2}$$

$\therefore \quad$ Solving for y

we get $y = 6\sqrt{2}$ m. (since body is going down negative value is chosen)

$\therefore \quad$ The location is $\left(0,-6\sqrt{2}m\right)$.

17.

$$v = \sqrt{\dfrac{GM}{r}} \qquad\qquad ........(1)$$

$$-\dfrac{GMm}{r} = -\dfrac{GMm}{R} + \dfrac{1}{2}mv'^2 \qquad\qquad .........(2)$$

From (1) and (2) we have $v' = v\sqrt{2\left(\dfrac{r}{R} - 1\right)}$

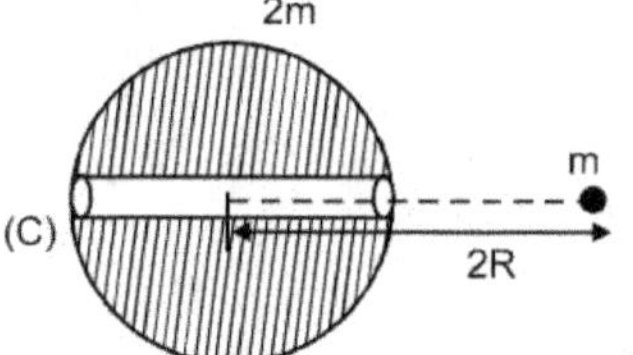


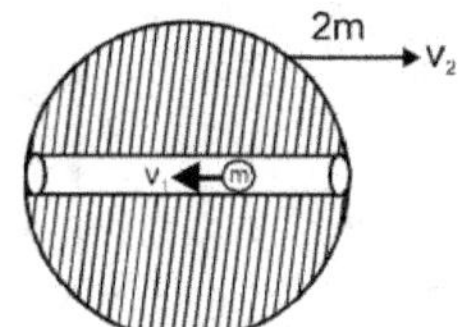


Applying momentum conservation,

$$0 = mv_1 - 2mv_2$$

$$\Rightarrow \quad v_2 = \frac{v_1}{2} \qquad \ldots\ldots\ldots\ldots(i)$$

From energy conservation,

$$k_i + U_i = k_f + U_f$$

$$0 + \left(-\frac{G(2m)}{2R}\right)m = \frac{1}{2}mv_1^2 + \frac{1}{2}(2m)v_2^2 + \left(-\frac{3}{2}\frac{G(2m)}{R}\right)(m) \qquad \ldots\ldots\ldots(ii)$$

Solving eqn.(i) & (ii) get,

$$v_1 = \sqrt{\frac{8Gm}{3R}}$$

(A) COM will be fixed so,

$$S_{cm} = \frac{m_1 s_1 + m_2 s_2}{m_1 + m_2}$$

$$0 = \frac{(m)(x) + (2m)(-(2R - x))}{m + 2m} \qquad \Rightarrow \qquad x = \frac{4R}{3}$$

(B) $F_{net} = 0 \quad \Rightarrow \quad a = 0$

(D) $W_{gr} = U\downarrow \quad \Rightarrow \quad W_{gr} = \left(-\frac{G(2m)}{2R}\right)m - \left(-\frac{3}{2}\frac{G(2m)}{R}\right)m .$

Let $x_0$ = extension in the spring when A is in equilibrium. Then,

$$k x_0 = \frac{1}{4\pi\epsilon_0}\frac{q^2}{r^2} \qquad \ldots\ldots (1)$$

Now let A be shifted by a small distance x towards B. Then the resultant force towards A is,

$$F_{res} = k(x_0 + x) - \frac{q^2}{4\pi\epsilon_0(r-x)^2} = k(x_0 + x) - \frac{q^2}{4\pi\epsilon_0 r^2}\left(1 - \frac{x}{r}\right)^{-2}$$

$$= k(x_0 + x) - \frac{q^2}{4\pi\epsilon_0 r^2}\left(1 + \frac{2x}{r}\right) ; \qquad x \ll r : \text{ Binomial expansion}$$

$$= kx - \frac{q^2}{2\pi\epsilon_0 r^3}x ; \text{ using (1)} \qquad F_{res} = \left(k - \frac{q^2}{2\pi\epsilon_0 r^3}\right)x$$

$$F \, \alpha \, x \therefore \text{ SHM with } T = 2\pi\sqrt{\frac{m}{k - \frac{q^2}{2\pi\epsilon_0 r^3}}} \quad \text{Ans.}$$

r real T, $\qquad k > \frac{q^2}{2\pi\epsilon_0 r^3} \qquad \therefore \qquad k_{min} = \frac{q^2}{2\pi\epsilon_0 r^3}$ Ans.

s. $T = 2\pi\sqrt{\frac{m}{k - \frac{q^2}{2\pi\epsilon_0 r^3}}}$ , $k_{min} = \frac{q^2}{2\pi\epsilon_0 r^3}$

**20.**

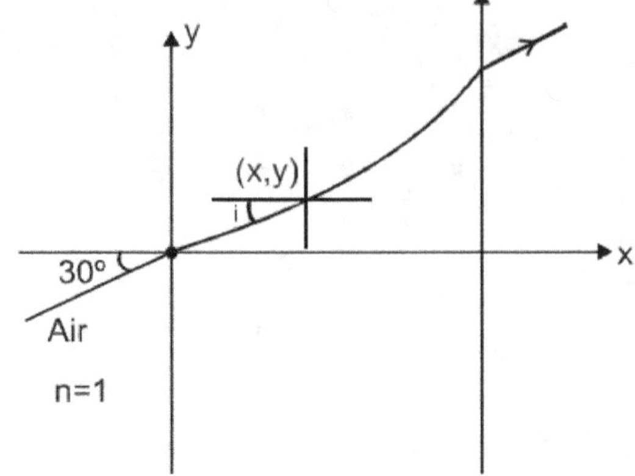


(a)     $1 \times \sin 30° = n \sin i$

$$\sin i = \frac{1}{2n}$$

$$\tan i = \frac{1}{\sqrt{4n^2 - 1}}$$



$$\frac{dy}{dx} = \frac{1}{\sqrt{x+3}}$$

$$\int_0^y dy = \int_0^x (x+3)^{-1/2}\, dx$$

$$y = 2\left(\sqrt{x+3} - \sqrt{3}\right)$$

(b)     when $x = 1$

$$y = 2(\sqrt{1+3} - \sqrt{3}), \qquad y = 2\left(2 - \sqrt{3}\right)$$

$\therefore$    Position at which ray comes out of the medium is $\left(1,\ 2(2 - \sqrt{3})\right)$.

(a) We can easily see that charge q is placed symetrically to surface ABCD, ABSR and ADQR. Charge q is also placed symetrically to rest of the surfaces.

If the flux through the surface ABCD is x and through RSPQ is y then the total flux will be 3x + 3y

Now by Gauss law

Now by Gauss law

$$\frac{q_{in}}{\varepsilon_0} = \phi$$

$$\Rightarrow \quad 3x + 3y = \frac{q}{\varepsilon_0}$$

$$\Rightarrow \quad x + y = \frac{q}{3\varepsilon_0}$$

(b) Flux through two surfaces are not same flux via ABCD is larger.

**Ans. (a) $\dfrac{q}{3\varepsilon_0}$ (b) Flux through two surfaces are not same flux via ABCD is larger.**

$$0 < x < a : V = \left[ -\int_0^x E_x dx \right] + V_{(0)} \qquad\qquad = 0 \ (\text{as } E_x = 0)$$

$$x > a \ ; \ V = -\int_a^x E_x dx + V_{(a)} \qquad = \left[ -\int_a^x \frac{\sigma}{\epsilon_0} dx \right] + V_{(a)} \qquad = \quad -\frac{\sigma}{\epsilon_0}(x - a)$$

$$x < 0 \ ; \ V = -\int_0^x E_x dx + V_{(0)} \qquad = -\left( -\frac{\sigma}{\epsilon_0}.x \right) + V_{(0)} \qquad = \quad \frac{\sigma}{\epsilon_0}.x \ .$$

Consider another identical hemisphere to complete a hollow spherical shell.
The potential at a point D due to half shell

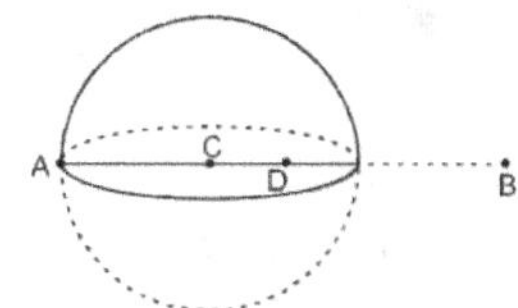


$$V_D = \frac{1}{2} \times \text{potential due to complete shell at D (due to symmetry)} = \frac{1}{2} \times \left( -\frac{G \cdot 2m}{R} \right) = -\frac{Gm}{R}$$

$$V_A = \frac{1}{2} \times \text{potential due to complete shell at A} = \frac{1}{2} \times \left( -\frac{G \cdot 2m}{R} \right) = -\frac{Gm}{R}$$

$$V_B = \frac{1}{2} \times \text{potential due to complete shell at B (again due to symmetry)} = \frac{1}{2} \times -\frac{G \times 2m}{2R} = -\frac{Gm}{2R}$$

**Ans.** $\quad V_A = V_D = -\dfrac{Gm}{R} , \ V_B = -\dfrac{Gm}{2R}$

**24.** Electric field inside the cavity $= \dfrac{\rho \vec{a}}{3\varepsilon_0}$ $\left[\begin{array}{l}\text{here } \vec{a} = \text{along line joining}\\ \text{Centers of sphere and cavity}\end{array}\right]$

Force on the electron inside the cavity $= \dfrac{\rho \vec{a}}{3\varepsilon_0}$ (e)

Cavity $\longrightarrow$ 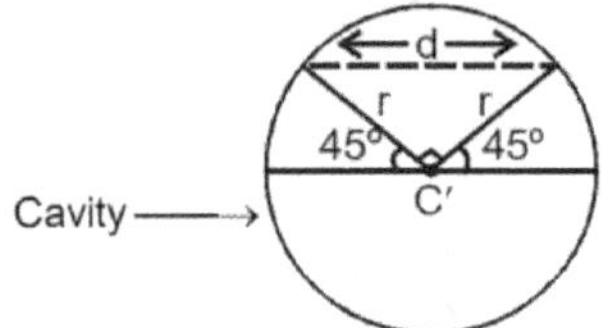

acceleration $= \dfrac{\rho a e}{3\varepsilon_0 m}.$

Now for distance $d = \sqrt{r^2 + r^2} = \sqrt{2}\,r$

Cavity $\longrightarrow$


by $S = ut + 1/2\, at^2$ , $\sqrt{2}\,r = \dfrac{1}{2} \times \dfrac{\rho a e}{3 m \varepsilon_0}\, t^2$ $\qquad \Rightarrow t = \left(\dfrac{6\sqrt{2}\, r m \varepsilon_0}{e a \rho}\right)^{\frac{1}{2}}$

**25.** Area covered by line joining planet and sun in time dt is

$$dS = \dfrac{1}{2} x^2 d\theta \quad ; \qquad \text{Areal velocity} = dS/dt = \dfrac{1}{2} x^2 d\theta /dt = \dfrac{1}{2} x^2 \omega$$

where   $x$ = distance between planet and sun
and    $\omega$ = angular speed of planet about sun.
From Keplers second law Areal velocity of planet is constant.

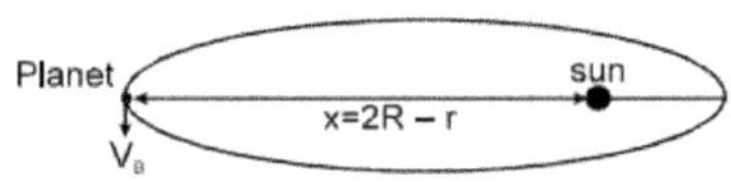


At farthest position

$$A = dS/dt = \dfrac{1}{2} (2R - r)^2 \omega = \dfrac{1}{2} (2R - r) [(2R - r)\, \omega] = \dfrac{1}{2} (2R - r) V_B$$

or    $V_B = \dfrac{2A}{2R - r}$ (least speed).  (Using values)

$V_B = 40$ km/s.

**26.** $R = 10\text{cm}$

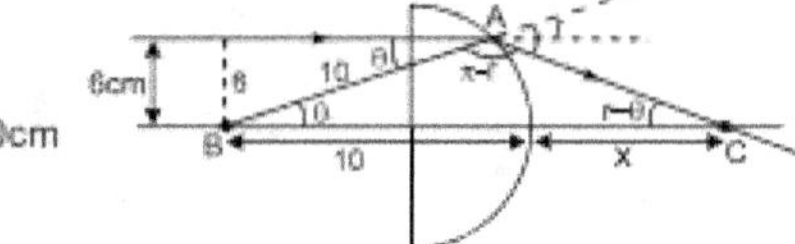

Applying snell's law $\quad \dfrac{\sin\theta}{\sin r} = \dfrac{3}{4} \quad \Rightarrow \quad r = 53°$

By sine law in $\triangle ABC \quad \dfrac{\sin(r-\theta)}{10} = \dfrac{\sin(\pi - r)}{(10+x)} \; ; \; \dfrac{10+x}{10} = \dfrac{4}{5\,(\sin r\cos\theta - \cos r\sin\theta)}$

$$= \dfrac{4}{5\left(\dfrac{4}{5}\times\dfrac{4}{5} - \dfrac{3}{5}\times\dfrac{3}{5}\right)} \; ; \; 10 + x = \dfrac{200}{7} \quad \Rightarrow \quad x = \dfrac{200-70}{7} = \dfrac{130}{7}$$

**27.** $a_1 = \dfrac{F}{m} = \dfrac{GM}{r^2}$

It is same in both cases

$\therefore \qquad \dfrac{a_1}{a_2} = 1$

**28.** $F = \displaystyle\int_{R}^{2R} \dfrac{GM\left(\dfrac{m}{R}\right)dx}{x^2} = \dfrac{GMm}{2R^2}$

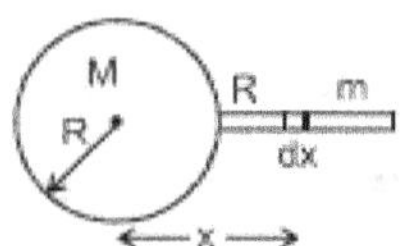

**29.** we have $f_1 = 50$ cm and $f_2 = 100$ cm

let the real distance between A and B be x. Also let refractive index of liquid be $\mu$. Then

$$\dfrac{1}{f_1} = \left(\dfrac{3}{2}-1\right)\left(\dfrac{1}{R_1} - \dfrac{1}{R_2}\right) \Rightarrow \left(\dfrac{1}{R_1} - \dfrac{1}{R_2}\right) = \dfrac{2}{f_1}$$

$$\dfrac{1}{f_1'} = \left(\dfrac{3}{2\mu}-1\right)\left(\dfrac{1}{R_1} - \dfrac{1}{R_2}\right) \Rightarrow \dfrac{1}{f_1'} = \dfrac{2}{f_1}\left(\dfrac{3-2\mu}{2\mu}\right)$$

$$\text{and } \dfrac{1}{f_2'} = \dfrac{2}{f_2}\left(\dfrac{3-2\mu}{2\mu}\right)$$

Now, for A we have

$$-\left(\dfrac{1}{200}\right) - \left(\dfrac{1}{-x}\right) = \dfrac{2}{50}\left(\dfrac{3-2\mu}{2\mu}\right)$$

$$\Rightarrow \qquad \dfrac{1}{x} = \dfrac{1}{200} + \dfrac{2}{50}\left(\dfrac{3-2\mu}{2\mu}\right) \qquad\qquad \text{...(1)}$$

Also for B we have

$$-\dfrac{1}{100} - \left(-\dfrac{1}{x}\right) = \dfrac{2}{100}\left(\dfrac{3-2\mu}{2\mu}\right)$$

$$\text{so,} \qquad \dfrac{1}{x} = \dfrac{1}{100} + \dfrac{2}{100}\left(\dfrac{3-2\mu}{2\mu}\right) \qquad\qquad \text{....(2)}$$

from (1) and (2) we get

$$\Rightarrow \quad \frac{2(3-2\mu)}{100\,(2\mu)} + \frac{1}{100} = \frac{1}{200} + \frac{2(3-2\mu)}{50\,(2\mu)}$$

$$\Rightarrow \quad \frac{2(3-2\mu)}{(2\mu)}\left[\frac{1}{50} - \frac{1}{100}\right] = \frac{1}{100} - \frac{1}{200} = \frac{1}{200}$$

$$\Rightarrow \quad \frac{(3-2\mu)}{2\mu} = \frac{1}{2} \qquad \Rightarrow 6 - 4\mu = \mu$$

$$\text{so } \mu = \frac{6}{5} = \frac{12}{10}$$

**30.** Image formation due to convex lens

$$\frac{1}{v} - \frac{1}{-36} = \frac{1}{30} \qquad \Rightarrow \quad v = \frac{30 \times 36}{6} = 180 \text{ cm}$$

This image will act like a virtual object for mirror and after reflection from mirror its image (shown by $I_2$) will be formed at 80 cm below optical axis of convex lens.

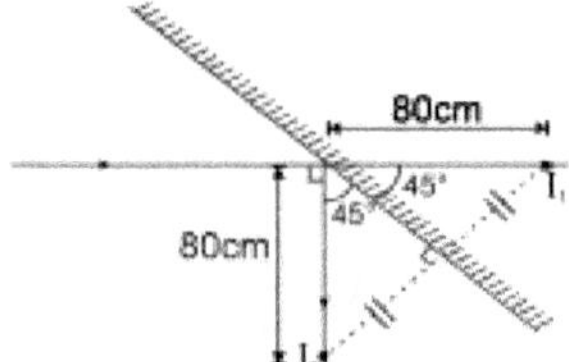


For concave lens, this image will be object at a position of 15 cm below the lens.
For final image formed by concave lens.

$$\frac{1}{20} - \frac{1}{15} = \frac{1}{f} \qquad \Rightarrow \quad \frac{1}{f} = -\frac{5}{300}$$

Also,

$$\frac{1}{f} = (\mu - 1)\left(-\frac{1}{R} - \frac{1}{R}\right)$$

or $\quad -\dfrac{5}{300} = \left(\dfrac{3}{2} - 1\right)\left(-\dfrac{2}{R}\right) \qquad \Rightarrow R = \dfrac{300}{5}$

R = 60 cm

**Ans. radius of curvature = 60 cm**

**31.**
$$\frac{GM}{(2R)^2} = \frac{GM'}{R^2}$$

$$\frac{M}{4} = M'$$

$$\frac{4}{3}\pi R^3 \rho_1 + \frac{4}{3}\pi(8R^3 - R^3)\rho_2 = 4\left(\frac{4}{3}\pi R^3 \cdot \rho_1\right)$$

$$\rho_1 + 7\rho_2 = 4\rho_1$$

$$\frac{\rho_1}{\rho_2} = \frac{7}{3}.$$

**.** $\delta = i + e - A$
$\delta_{min} = 60°$ when $i = e$
$\therefore 60° = 2i - A = 2(60°) - A \qquad\qquad \therefore A = 60°$

$$\therefore \mu = \frac{\sin\left(\dfrac{A + \delta_{min}}{2}\right)}{\sin\left(\dfrac{A}{2}\right)} = \frac{\sin\left(\dfrac{60 + 60}{2}\right)}{\sin\left(\dfrac{60}{2}\right)} = \sqrt{3}$$

**.** When angle of incidence is $i_1$, $e = 40°$
(from reversibility of ray)
also $\delta = 70°$
$\therefore 70° = i_1 + 40° - A$
$\therefore i_1 = 90°$

**.** $\vec{E} = \dfrac{kQ}{x^2}$

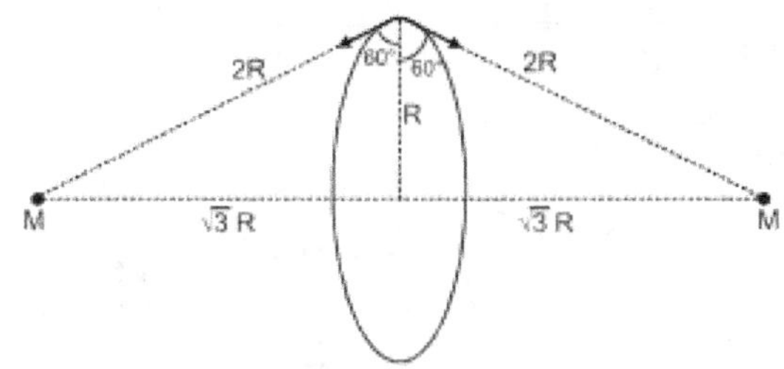

$$\vec{E}_1 = \frac{1}{4\pi\varepsilon_0}\ \frac{\frac{4}{3}\pi x^3 \rho}{x^2} = \frac{\rho d}{3\varepsilon_0}(d - x)$$

$$E_{net} = E_1 + E_2 = \frac{\rho(d - x)}{3\varepsilon_0} + \frac{\rho x}{3\varepsilon_0}$$

$$E = \frac{\rho d}{3\varepsilon_0}$$

**.** $V = -\int E - dx$

$$\int_{V_1}^{V_2} V = -\int_0^d \frac{\rho d}{3\varepsilon_0}\, dx\ ;\ \ V_2 - V_1 = -\frac{\rho d^2}{3\varepsilon_0}\ ;\ |\Delta V| = \frac{\rho d^2}{3\varepsilon_0}$$

**' to 39.**

**ɔl.**

$$F_{net} = 2\left(\frac{GMm}{4R^2}\right)\cos 60° = \frac{GMm}{4R^2}$$

$$F_{net} = \frac{GMm}{4R^2} = \frac{mv^2}{R} \quad\Rightarrow\quad v = \sqrt{\frac{GM}{4R}} = \frac{1}{2}\sqrt{\frac{GM}{R}}$$

$$T = \frac{2\pi R}{v} = \frac{4\pi R^{3/2}}{\sqrt{GM}}$$

Average force on planet in half revolution.

$$F_{avg} = \frac{2mv}{T/2} = \frac{4mv}{T} = \frac{4mv}{\frac{2\pi R}{v}} = \frac{2mv^2}{\pi R} = \frac{GMm}{2\pi R^2}$$

**40 to 42.**

Potentials at the centre

$$v_1 = \frac{1}{4\pi\varepsilon_0}\frac{q}{r}; \qquad v_2 = \frac{1}{4\pi\varepsilon_0}\frac{q}{r}$$

Potential energy in situation I is

$$U_1 = 3 \times \frac{1}{4\pi\varepsilon_0}\frac{(q/3)^2}{(\sqrt{3}R)} = \frac{1}{12\sqrt{3}\pi\varepsilon_0}\frac{q^2}{R}$$

When one charge is removed, the field intensity at the centre is due to the removed charge only.

$$E_1 = \frac{1}{4\pi\varepsilon_0}\frac{q/3}{r^2}$$

$$E_2 = \frac{1}{4\pi\varepsilon_0}\frac{q/4}{r^2} \qquad \therefore \frac{E_1}{E_2} = \frac{4}{3}$$

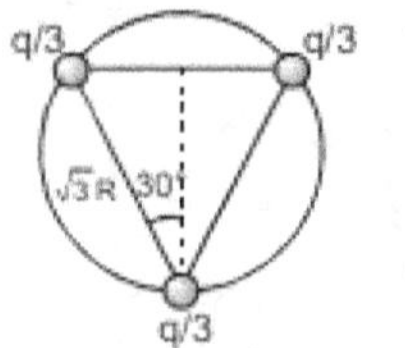


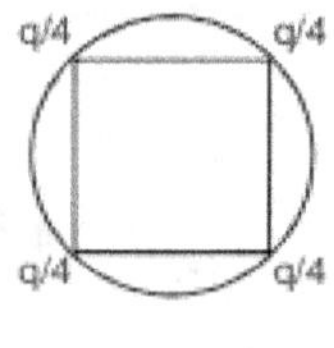


**43.**

$$C = \sin^{-1}\left(\frac{1}{2/1}\right) = 30°$$

for    $i = 37$, TIR    so, $\delta = \pi - 2\,(37°) = 104°$

$i = 25$,   Refraction    $\delta < \frac{\pi}{2} - C$

$i = 45°$, TIR    so, $\delta = \pi - 2\left(\frac{\pi}{4}\right) = 90°$

By applying snells law for prism :

$$i = 90,$$
$$r_1 = 30, \qquad r_2 = 30$$
$$e = 45$$
$$\delta = 90 + 45 - 60 = 75°$$

**44.**

(A) Electrostatic potential energy $= \dfrac{1}{4\pi\varepsilon_0}\dfrac{(-Q)^2}{2a} = \dfrac{Q^2}{8\pi\varepsilon_0 a}$

(B) Electrostatic potential energy $= \dfrac{1}{4\pi\varepsilon_0}\left[\dfrac{(-Q)\times(-Q)}{5a/2} + \dfrac{(-Q)^2}{2(5a/2)}\right] = \dfrac{3}{20}\dfrac{Q^2}{\pi\varepsilon_0 a}$

(C) Electrostatic potential energy $= \dfrac{1}{4\pi\varepsilon_0}\dfrac{3Q^2}{5a} = \dfrac{3}{20}\dfrac{Q^2}{\pi\varepsilon_0 a}$

(D) Electrostatic potential energy $= \dfrac{1}{4\pi\varepsilon_0}\left[\dfrac{3Q^2}{5a} + \dfrac{(-Q)^2}{2(2a)} + \dfrac{(-Q)\times(-Q)}{2a}\right] = \dfrac{27Q^2}{80\pi\varepsilon_0 a}$

**45.**

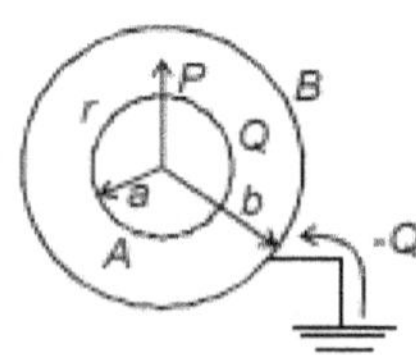


Field at P is only due to A $= \dfrac{kQ}{(2)^2} = \dfrac{kQ}{4}$

Potential at P $= V_{\text{due to A}} + V_{\text{due to B}} = \dfrac{kQ}{2} - \dfrac{kQ}{3}$

Electric field outside B is due to 'A's induced charge on B + A's charge = zero.

# Mock Test – 2

A cannon fires successively two shells from the same point with velocity $v_0 = 250$ m/s ; the first at the angle $\theta_1 = 60°$ and the second at the angle $\theta_2 = 45°$ to the horizontal, the azimuth being the same. Neglecting the air drag, find the approximate time interval between firings leading to the collision of the shells ($g = 9.8$ m/s$^2$.)

(A) 11 sec  (B) 6 sec  (C) 15 sec  (D) 5 sec

Each of the two block shown in the figure has mass m. The pulley is smooth and the coefficient of friction for all surfaces in contact is $\mu$. A constant horizontal force P applied in two cases shown in such a way that block A start just sliding then the value of minimum force P in case-I and case-II is :

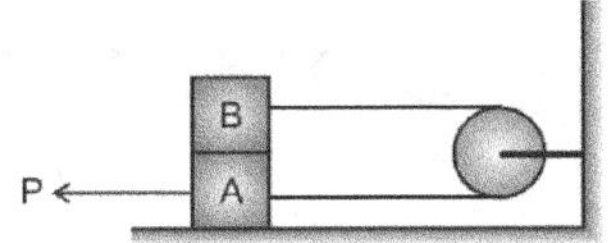

(A) $2\mu\,mg$, $3\mu\,mg$   (B) $3\mu\,mg$, $2\mu\,mg$
(C) $4\mu\,mg$, $3\mu\,mg$   (D) $3\mu\,mg$, $3\mu\,mg$

3. A particle is projected with speed 30m/s at angle 22.5° with horizontal from ground as shown. AB and CD are parallel to y-axis and B is highest point of trajectory of particle. CD/AB is

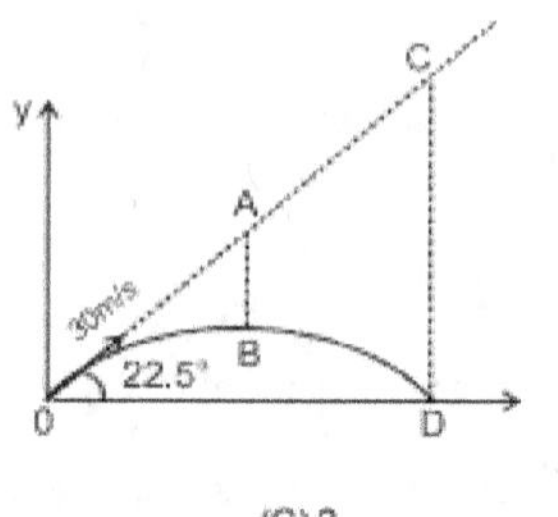


(A) 3         (B) 3/2         (C) 2         (D) 4

4. A block of mass m is pulled on an incline surface having coefficient of friction $\mu = 1$ & angle of inclination $\theta = 30°$, with the horizontal, such that required external force is minimum. The angle made by this force with the incline is :

(A) 45°         (B) 30°         (C) 75°         (D) 53°

5. Two cars A and B moving on two straight tracks inclined at an angle 60° heading towards the crossing initially their positions are as shown in the figure. Both cars have same speed. Minimum seperation between them during their motion will be.

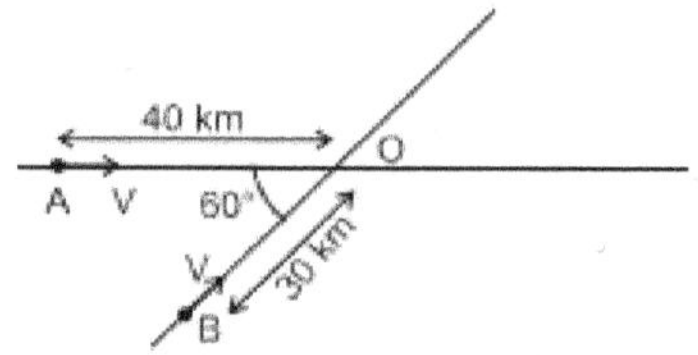


(A) 10 km         (B) $5\sqrt{3}$ km         (C) 5 km.         (D) $\dfrac{20}{\sqrt{3}}$ km

6. Three particles A, B and C situated at vertices of an equilateral triangle, all moving with same constant speed such that A always move towards B, B always towards C and C always towards A. Initial seperation between each of the particle is a. O is the centroid of the triangle. Distance covered by particle A when it completes one revolution around O is

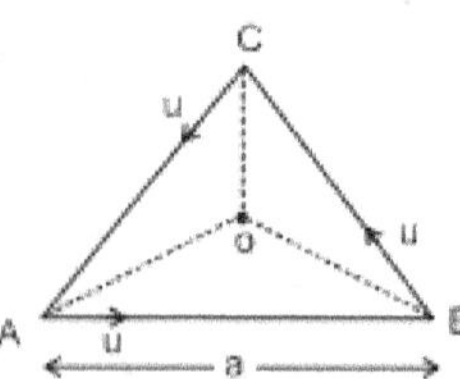


(A) $2a\left(1-e^{-2\sqrt{3}\pi}\right)$    (B) $\dfrac{2a}{3}\left(1-e^{-2\sqrt{3}\pi}\right)$    (C) $a\left(1+e^{-2\sqrt{3}\pi}\right)$    (D) $\dfrac{2a}{3}\left(1-e^{-\sqrt{3}\pi}\right)$

ABC is a triangle in vertical plane. Its two base angles $\angle BAC$ and $\angle BCA$ are $45°$ and $\tan^{-1}(1/3)$ respectively. A particle is projected from point A such that it passes through vertices B and C. Find angle of projection in degrees:

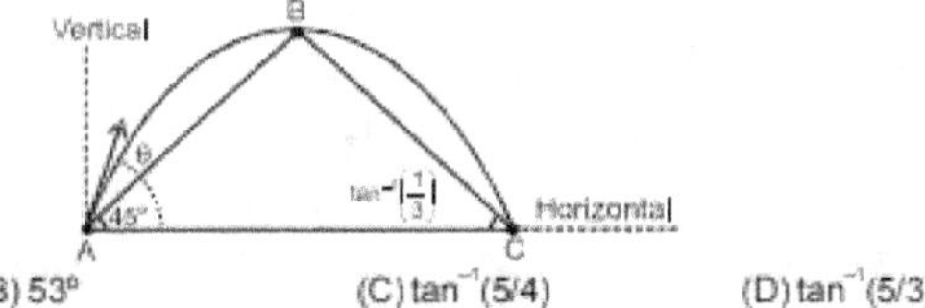


(A) $60°$      (B) $53°$      (C) $\tan^{-1}(5/4)$      (D) $\tan^{-1}(5/3)$

A rope of negligible mass passes over a pulley of negligible mass attached to the ceiling, as shown in figure. One end of the rope is held by Student A of mass 70 kg, who is at rest on the floor. The opposite end of the rope is held by Student B of mass 60 kg, who is suspended at rest above the floor. The minimum acceleration $a_0$ with which the Student B should climb up the rope to lift the Student A upward off the floor. ($g = 10$ m/s$^2$)

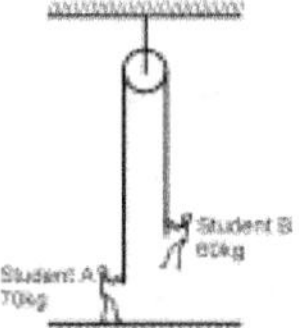


(A) $\frac{1}{3}$ m/s$^2$      (B) $\frac{2}{3}$ m/s$^2$      (C) $\frac{4}{3}$ m/s$^2$      (D) $\frac{5}{3}$ m/s$^2$

A balloon is tied to a block. The mass of the block is 2kg. The tension of the string between the balloon and the block is 30N. Due to the wind, the string has an angle $\theta$ relative to the vertical direction. $\cos\theta = 4/5$ and $\sin\theta = 3/5$. Assume the acceleration due to gravity is $g = 10$ m/s$^2$. Also assume the block is small so the force on the block from the wind can be ignored. Then the x-component and the y-component of the acceleration a of the block.

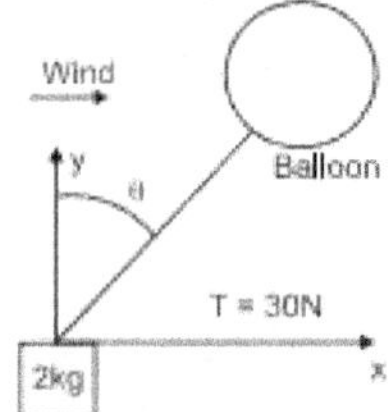


(A) 9 m/s$^2$, 2 m/s$^2$      (B) 9 m/s$^2$, 12 m/s$^2$      (C) 18 m/s$^2$, 2 m/s$^2$      (D) 18 m/s$^2$, 12 m/s$^2$

The maximum angle to the horizontal at which a stone can be thrown so that it always moves away from the thrower will be :

(A) $\sin^{-1}\left(\frac{\sqrt{2}}{3}\right)$      (B) $\sin^{-1}\left(\frac{2\sqrt{2}}{3}\right)$      (C) $\sin^{-1}\left(\frac{1}{\sqrt{3}}\right)$      (D) $\sin^{-1}\left(\sqrt{\frac{2}{3}}\right)$

A man starts walking on a circular track of radius R. First half of the distance he walks with speed $V_1$, half of the remaining distance with speed $V_2$, then half of the remaining time with $V_1$ and rest with $V_2$ and completes the circle. Average speed of the man during entire motion in which he completes the circle is.

(A) $\dfrac{2V_1V_2(V_1+V_2)}{V_2^2+2V_1^2+2V_1V_2}$    (B) $\dfrac{4V_1V_2(V_1+V_2)}{V_1^2+2V_2^2+5V_1V_2}$    (C) $\dfrac{V_1V_2(V_1+2V_2)}{V_1^2+V_2^2+4V_1V_2}$    (D) $\dfrac{(V_1+2V_2)^2}{V_1+V_2+2V_1^2V_2^2}$

12. Two blocks of masses 8kg and 6kg are connected with a string & placed on a rough horizontal surface. Surface itself is accelerating up with constant acceleration 2 m/s². Two forces 60 N each are acting on the two blocks as shown. Friction coefficient for 8kg is 0.5 & that for 6 kg is 0.6. Tension in the string is : (g = 10 m/s²)

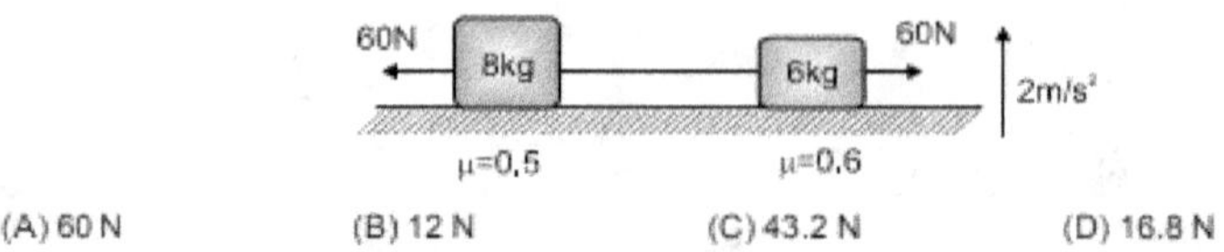


(A) 60 N      (B) 12 N      (C) 43.2 N      (D) 16.8 N

13. Block A of weight 500 N and block B of weight 700 N are connected by rope pulley system as shown. The largest weight C that can be suspended without moving block A and B is W. The coefficient of friction for all plane surfaces of contact is 0.3. The pulleys are ideal. Find $\dfrac{W}{90}$.

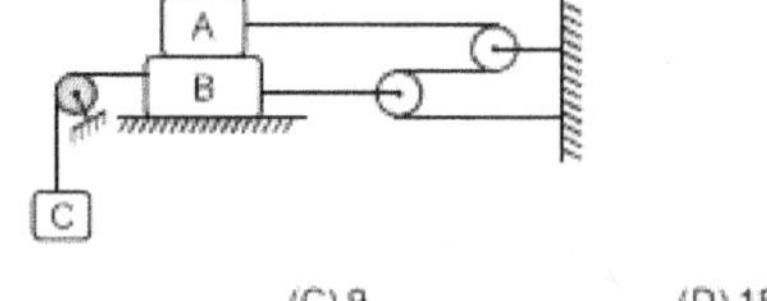


(A) 12      (B) 22      (C) 9      (D) 18

14. O is a point at the bottom of a rough plane inclined at an angle $\alpha$ to the horizontal. Coefficient of friction between AB is $\dfrac{\tan\alpha}{2}$ and between BO is $\dfrac{3\tan\alpha}{2}$. B is the mid-point of AO. A block is released from rest at A. Then identify which graphs are correct during motion of block from point A to O taking direction down the incline plane as positive ($\sin\alpha = 1/5$) :

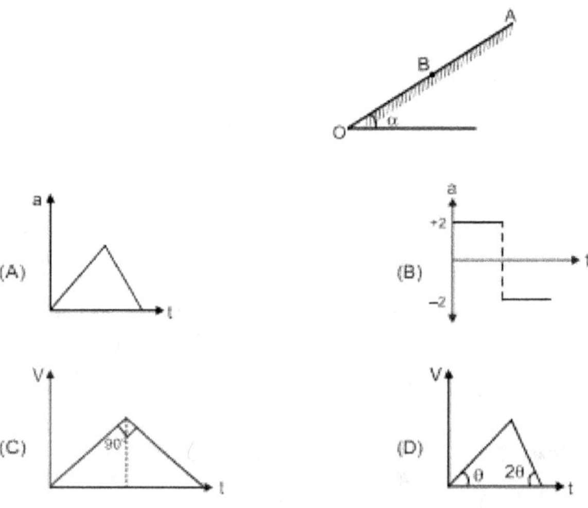

A block D of mass 10 kg is placed on smooth horizontal surface over it another block A of same mass is placed. A horizontal force F is applied on block B.

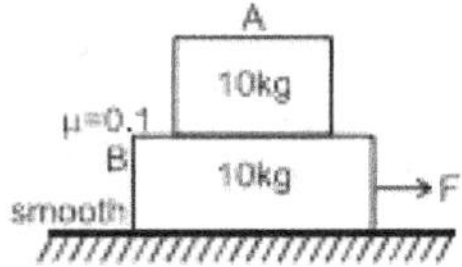


$S_1$ : No block will move unless F > 10 N.
$S_2$ : Block A will move towards left.
$S_3$ : Acceleration of block B will never be less than that of A.
$S_4$ : The relative motion between A and B will start when F exceeds 10 N.

(A) F F F F

(B) T T T T

(C) F F T F

(D) T T F F

Block A of mass m is placed on a plank B. A light support S is fixed on plank B and is attached with the block A with a spring of spring constant K. Consider that initially spring is in its natural length. If the plank B is given

an acceleration a, then maximum compression in the spring is $\dfrac{xma}{k}$. Find the value of x. (All the surfaces are

smooth)

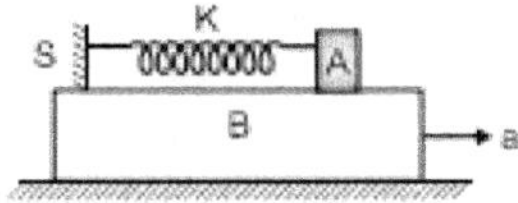


(A) $\dfrac{ma}{2k}$

(B) $\dfrac{2ma}{k}$

(C) $\dfrac{ma}{k}$

(D) $\dfrac{4ma}{k}$

Mass m shown in figure is in equilibrium. If it is displaced further by x and released find its acceleration just after it is released. Take pulleys to be light & smooth and strings light.

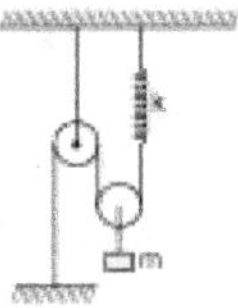


(A) $\dfrac{4kx}{5m}$

(B) $\dfrac{2kx}{5m}$

(C) $\dfrac{4kx}{m}$

(D) none of these

18. Both the blocks shown in figure have same mass 'm'. All the pulley and strings are massless.

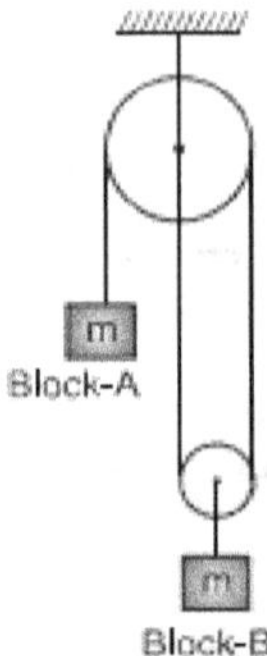


(A) Acceleration of block A is $\dfrac{2g}{5}$

(B) Acceleration of block A is $\dfrac{g}{5}$

(C) Acceleration of block B is $\dfrac{g}{5}$

(D) Tension in the string attached with A is $\dfrac{3mg}{5}$

19. Two cars $C_1$ & $C_2$ are moving in parallel lanes in the same direction at speeds 90 kph & 108 kph respectively. (see figure). As the traffic signal turns red , both applies brake (assume constant retardation) simultaneously. If they both stop together at the dead line :

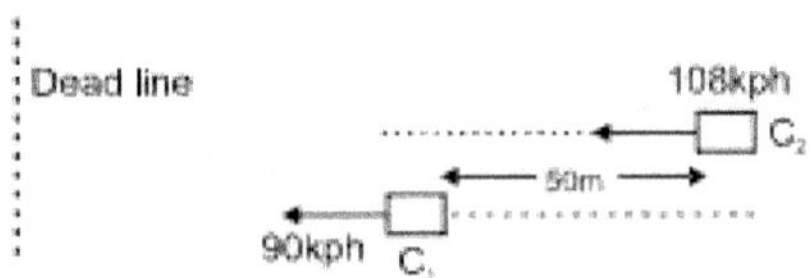


(A) distance of dead line from $C_2$ is 300 m
(B) distance of dead line from $C_1$ is 250 m
(C) time taken by the cars to ends up after the signal turn red is 15 sec
(D) time taken by the cars to ends up after the signal turn red is 20 sec

20. A man is standing on a road and observes that rain is falling at angle $45^\circ$ with the vertical. The man starts running on the road with constant acceleration $0.5 \ m/s^2$. After a certain time from the start of the motion, it appears to him that rain is still falling at angle $45^\circ$ with the vertical, with speed $2\sqrt{2}$ m/s . Motion of the man is in the same vertical plane in which the rain is falling. Then which of the following statement(s) are true.
(A) It is not possible
(B) Speed of the rain relative to the ground is 2 m/s.
(C) Speed of the man when he finds rain to be falling at angle $45^\circ$ with the vertical, is 4m/s.
(D) The man has travelled a distance 16m on the road by the time he again finds rain to be falling at angle $45^\circ$.

Two blocks A and B of equal mass m are connected through a massless string and arranged as shown in figure. The wedge is fixed on horizontal surface. Friction is absent everywhere. When the system is released from rest.

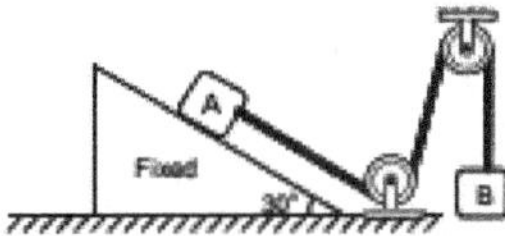


(A) tension in string is $\dfrac{mg}{2}$

(B) tension in string is $\dfrac{mg}{4}$

(C) acceleration of A is $\dfrac{g}{2}$

(D) acceleration of A is $\dfrac{3}{4}g$

In the figure shown, A & B are free to move. All the surfaces are smooth. $(0 < \theta < 90°)$

(A) the acceleration of A will be more than $g \sin \theta$

(B) the acceleration of A will be less than $g \sin \theta$

(C) normal force on A due to B will be more than $mg \cos \theta$

(D) normal force on A due to B will be less than $mg \cos \theta$

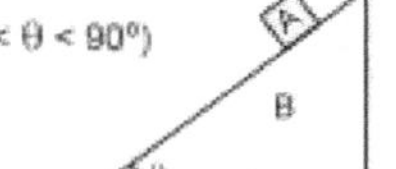


In given arrangement, 10 kg and 20 kg blocks are kept at rest on two fixed inclined planes. All strings and pulleys are ideal. value(s) of m for which system remain in equilibrium are: $(g = 10 \ m/s^2)$

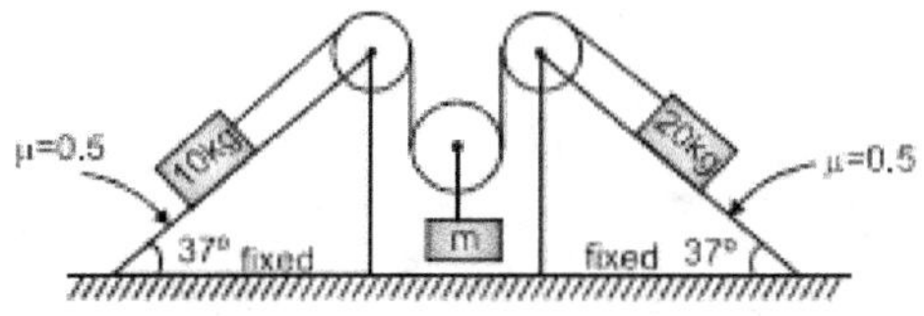


(A) m = 6 kg

(B) m = 13 kg

(C) m = 9 kg

(D) m = 12 kg

The system shown is in limiting equilibrium. The coefficient of friction for all contact surfaces is $\dfrac{1}{4}$.

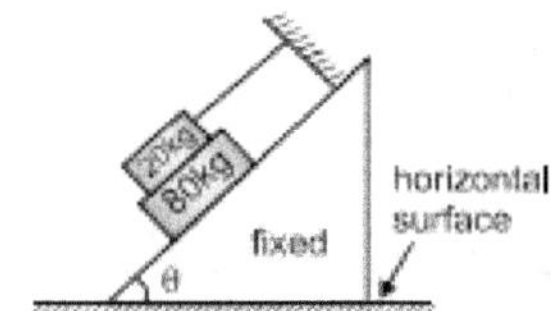


(A) $\tan\theta = \dfrac{3}{8}$

(B) Tension in the string $= \left(\dfrac{100}{3}g\sin\theta\right) N$

(C) Net frictional force on 80 kg block is $(80 \ g \sin\theta)N$

(D) Force exerted by 20 kg block on 80 kg block is $(20 \ g \cos\theta)$

25. A small body is projected with a velocity of 20.5 ms$^{-1}$ along rough horizontal surface. The coefficient of friction ($\mu$) between the body and surface changes with time t (in s) as the body moves along the surface. Find the velocity at the end of 4s in m/s

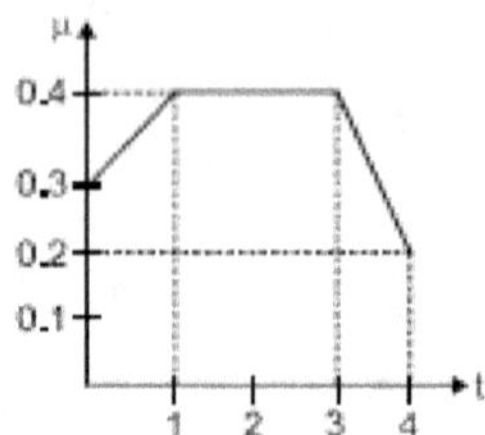


26. Position (in m) of a particle moving on a straight line varies with time (in sec) as $x = t^3/3 - 3t^2 + 8t + 4$ (m). Consider the motion of the particle from t = 0 to t = 5 sec. $S_1$ is the total distance travelled and $s_2$ is the distance travelled during retardation. If $s_1/s_2 = \dfrac{(3\alpha + 2)}{11}$ the find $\alpha$.

27. A block of 7 kg is placed on a rough horizontal surface and is pulled through a variable force F(in N) = 5t, where 't' is time in second, at an angle of 37° with the horizontal as shown in figure. The coefficient of static friction of the block with the surface is one. If the force starts acting at t = 0 s, Find the time at which the block starts to slide. (Take g = 10 m/s$^2$) :

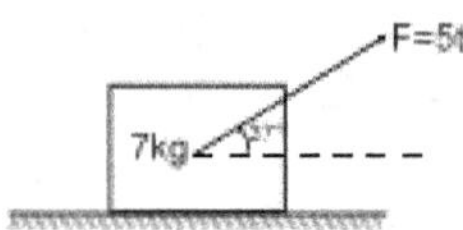


28. A block of mass $m_1$ is placed on a wedge of an angle $\theta$, as shown. The block is moving over the inclined surface of the wedge. Friction coefficient between the block and the wedge is $\mu_1$, whereas it is $\mu_2$ between the wedge and the horizontal surface. If $\mu_1 = \dfrac{1}{2}$, $\theta = 45°$, $m_1 = 4$ kg, $m_2 = 5$kg and g = 10 m/s$^2$, find minimum value of $\mu_2$ so that the wedge remains stationary on the surface. Express your answer in multiple of $10^{-3}$.

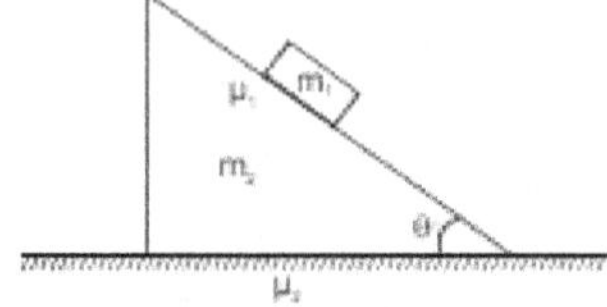


29. One has to throw a particle from one side of a fixed sphere, in diametrical plane to another side such that it just grazes the sphere. Minimum possible speed for this is $\sqrt{2gR(\sqrt{\alpha} + \sqrt{\beta})}$. Find $\alpha + \beta$.

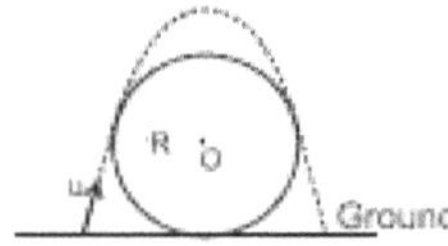

In the situation shown coefficient of friction between A and B is 0.5 and between B and C is 0.3. Friction acting between B and C is xN then $\dfrac{9x}{7}$ is :

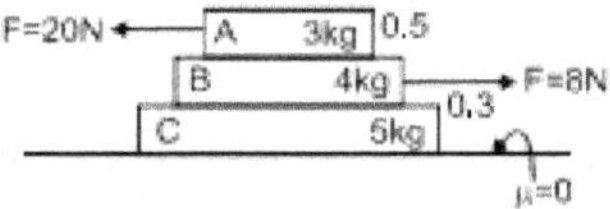


In the figure acceleration of bodies A, B and C are shown with directions. Values b and c are w.r.t. ground. Whereas a is acceleration of block A w.r.t. wedge C. Acceleration of block A w.r.t. ground is $\sqrt{\beta}$ m/s². Find β.
(Use b = c = 1 m/s², θ = 60°)

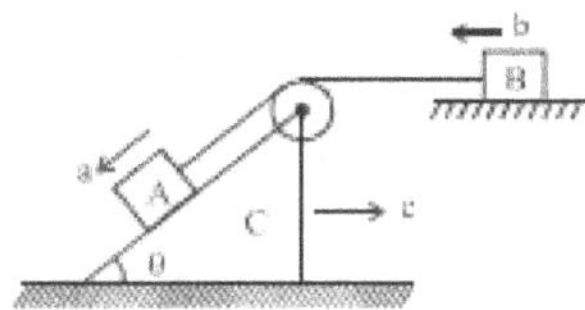


## mprehension-1

Two blocks A and B of masses m and 2m are initially at rest. Length of block B is L and the block A is placed at the right end corner of block B and the friction coefficient between them is μ = 1/2. At t = 0 a constant force F $= \dfrac{5mg}{2}$ begins to act on block B towards right. Just when the block A leaves B, wind begins to blow along y - direction which exerts a constant force $\dfrac{mg}{2}$ on A. Assume the size of block A is small compared to B and neglect any rotational effects and toppling of block B. (Given h = 1/2 m, L = 1m and g = 10 m/s²)

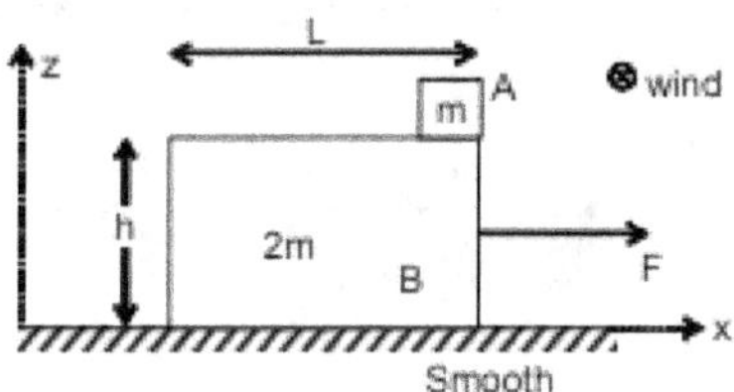


Find ratio of the displacements of block A along x and y directions $S_x/S_y$ after the time block A leaves the surface of B till the time it reaches ground

(A) $\dfrac{1}{2}$     (B) $\dfrac{1}{4}$     (C) 4     (D) $\sqrt{\dfrac{8}{5}}$

33. The magnitude of relative acceleration of A with respect to B (in m/s$^2$) just after the block A leaves B is (assume wind does not effects motion of B)

(A) $\sqrt{10}g$    (B) $\dfrac{\sqrt{29}g}{4}$    (C) $\dfrac{9\sqrt{5}}{4}$    (D) $\dfrac{3\sqrt{5}}{4}g$

**Comprehension : 2**

A smooth wedge of mass M is pulled towards left with an acceleration a = $g\cot\theta$ on a horizontal surface and a block of mass m is released w.r.t wedge. Then answer the following :

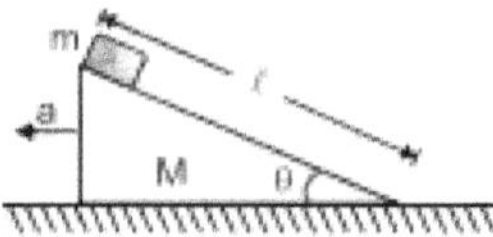

34. Time taken by the block to reach the ground is :

(A) $\sqrt{\dfrac{2\ell\sin\theta}{g}}$    (B) $\sqrt{\dfrac{2\ell}{g\sin\theta}}$    (C) $\sqrt{\dfrac{2\ell}{g\cos\theta}}$    (D) $\dfrac{v}{g\sin\theta}$

35. Normal reaction between the wedge and block is :
(A) mg cos$\theta$    (B) mg sec$\theta$    (C) mg cot$\theta$    (D) zero

36. Normal reaction offered by ground to the wedge is :
(A) M g    (B) (M + m)g cot$\theta$    (C) mg sin$^2\theta$ + Mg    (D) (M + m)g

**Comprehension : 3**

Three identical uniform blocks of mass m each and length L are placed on a smooth fixed horizontal surface as shown. There is friction between A and B (Friction cofficient $\mu$) while there is no friction between A and C.

At the instant shown, that is at t = 0; the block A has horizontal velocity of magnitude u towards right, whereas speed of B and C is zero. At the instant block A has covered a distance L relative to block B velocity of all blocks are same.

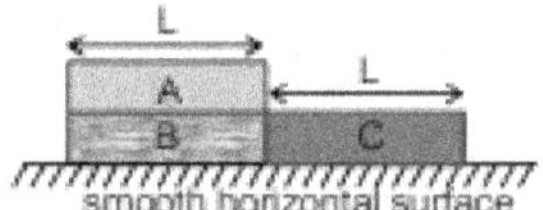

37. The speed of block A when it just looses contact with B is :

(A) $\dfrac{u}{2}$    (B) $\dfrac{u}{3}$    (C) $\dfrac{u}{4}$    (D) $\dfrac{2u}{3}$

38. The magnitude of total work done by friction on system of three blocks is :

(A) $-\dfrac{1}{3}mu^2$    (B) $-\dfrac{1}{4}mu^2$    (C) $-\dfrac{2}{3}mu^2$    (D) $\dfrac{1}{3}mu^2$

39. The value of $\mu$ is

(A) $\mu = \dfrac{3}{2}\dfrac{u^2}{gL}$    (B) $\mu = \dfrac{1}{2}\dfrac{u^2}{gL}$    (C) $\mu = \dfrac{u^2}{gL}$    (D) $\mu = \dfrac{2}{3}\dfrac{u^2}{gL}$

Initially both blocks are at rest on a horizontal surface and string is just tight. At t = 0, two constant horizontal forces $F_1$ and $F_2$ start acting on blocks as shown. $f_1$ and $f_2$ are friction forces acting on 10 kg and 20 kg block (co–efficient of friction between blocks and ground are 0.5). Values of $F_1$ and $F_2$ are given in column–I. Then match magnitudes of $f_1$, $f_2$ and direction of $\vec{f_1}$ with corresponding values of $F_1$ and $F_2$ given in column–I [g = 10 m/s²].

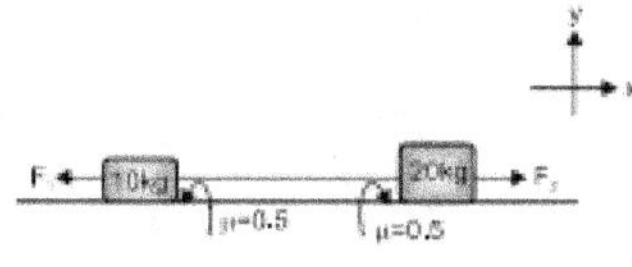


<table>
<tr><td colspan="2" align="center">Column–I</td><td colspan="2" align="center">Column–II</td></tr>
</table>

| Column–I | Column–II |
|---|---|
| (A) $F_2 = 120$ N, $F_1 = 40$ N | (p) $f_2 = 100$ N , $f_1 = 20$ N |
| (B) $F_2 = 160$ N, $F_1 = 40$ N | (q) $f_2 = 20$ N , $f_1 = 50$ N |
| (C) $F_2 = 60$ N, $F_1 = 90$ N | (r) $f_2 = 70$ N , $f_1 = 50$ N |
| (D) $F_2 = 20$ N, $F_1 = 90$ N | (s) unit vector in direction of $\vec{f_1}$ is $\hat{i}$ |
| | (t) unit vector in direction of $\vec{f_1}$ is $-\hat{i}$ |

A square platform of side length 8 m is situated in x–z plane such that it is at 16 m from the x–axis and 8 m from the z-axis as shown in figure. A particle is projected with velocity $\vec{v} = (v_2\hat{i} + 25\hat{j})$ m/s relative to wind from origin and at the same instant the platform starts with acceleration $\vec{a} = (2\hat{i} + 2.5\hat{j})$ m/s². Wind is blowing with velocity $v_1\hat{k}$ . (g = 10 m/s²)

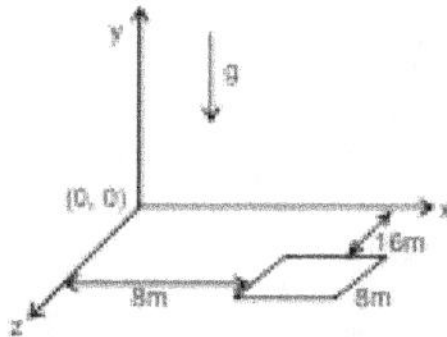


| | List I | | List II |
|---|---|---|---|
| (P) | Least possible values of $v_2$ (in m/s) so that particle hits the platform or edge of platform is | (1) | 4 |
| (Q) | Least possible value of $v_1$ (in m/s) so that particle hits the platform or edge of platform is | (2) | 6 |
| (R) | If t is the time (in second) after particle hits the platform then 2t is equal to | (3) | 8 |
| (S) | Value of displacement with respect to ground (in m) of the particle in y–direction, when $v_2$ has its minimum possible value is (till particle hits the platform or edge of platform) | (4) | 20 |

**Codes :**

| | P | Q | R | S |
|---|---|---|---|---|
| (A) | 2 | 4 | 3 | 1 |
| (B) | 2 | 1 | 3 | 4 |
| (C) | 2 | 3 | 4 | 1 |
| (D) | 2 | 1 | 4 | 3 |

42. **Match the following :**

Three blocks of masses $m_1$, $m_2$ and M are arranged as shown in figure. All the surfaces are frictionless and string is inextensible. Pulleys are light. A constant force F is applied on block of mass $m_1$. Pulleys and string are light. Part of the string connecting both pulleys is vertical and part of the strings connecting pulleys with masses $m_1$ and $m_2$ are horizontal.

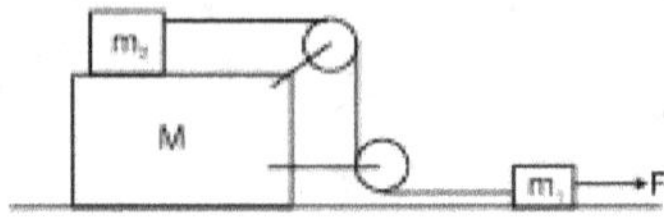



(P) Acceleration of mass $m_1$      (1) $\dfrac{F}{m_1}$

(Q) Acceleration of mass $m_2$      (2) $\dfrac{F}{m_1 + m_2}$

(R) Acceleration of mass M      (3) zero

(S) Tension in the string      (4) $\dfrac{m_2 F}{m_1 + m_2}$

|     | P | Q | R | S |
| --- | --- | --- | --- | --- |
| (A) | 2 | 2 | 3 | 4 |
| (B) | 2 | 1 | 3 | 4 |
| (C) | 2 | 2 | 4 | 1 |
| (D) | 2 | 1 | 3 | 1 |

# Answers

| | | | | | | | | | | | | | |
| --- | --- | --- | --- | --- | --- | --- | --- | --- | --- | --- | --- | --- | --- |
| **1.** | (B) | **2.** | (B) | **3.** | (C) | **4.** | (D) | **5.** | (B) | **6.** | (C) | **7.** | (A) |
| **8.** | (B) | **9.** | (C) | **10.** | (C) | **11.** | (B) | **12.** | (A) | **13.** | (B) | **14.** | (D) |
| **15.** | (B) | **16.** | (B) | **17.** | (B) | **18.** | (A,B,C) | **19.** | (A,C) | **20.** | (B,C) | **21.** | (A,D) |
| **22.** | (A,B,D) | **23.** | (A,B,C,D) | **24.** | 6 | **25.** | 40 | **26.** | 65 | **27.** | 1 | | |
| **28.** | 6 | **29.** | 12 | **30.** | 60 | **31.** | 07 | **32.** | 6 | **33.** | (A) | **34.** | (D) |
| **35.** | (C) | **36.** | (A) | **37.** | (B) | **38.** | (C) | **39.** | (A) | **40.** | (B) | **41.** | (C) |
| **42.** | (D) | **43.** | (B) | **44.** | (A) | **45.** | (D) | | | | | | |

# Solution

For particle -1 $\quad y = \sqrt{3}\,x - \dfrac{gx^2}{2u^2(1/4)} \Rightarrow y = \sqrt{3}\,x - \dfrac{2gx^2}{u^2}$

For particle-2 $\quad y = x - \dfrac{gx^2}{2u^2(1/2)} \Rightarrow y = x - \dfrac{gx^2}{u^2}$

$x - \dfrac{gx^2}{u^2} = \sqrt{3}\,x - \dfrac{2gx^2}{u^2}$

$x(\sqrt{3} - 1) = \dfrac{gx^2}{u^2} \quad \Rightarrow \quad x = \dfrac{u^2}{g}(\sqrt{3} - 1)$

for particle -1

$$u(1/2)\,t_1 = \dfrac{u^2}{g}(\sqrt{3} - 1) \Rightarrow t_1 = \dfrac{2u}{g}(\sqrt{3} - 1)$$

$$u(1/\sqrt{2})\,t_2 = \dfrac{u^2}{g}(\sqrt{3} - 1) \Rightarrow t_2 = \dfrac{\sqrt{2}u}{g}(\sqrt{3} - 1)$$

$$\Delta t = u/g\ (2 - \sqrt{2})(\sqrt{3} - 1) = 10.9 \text{ sec} \approx 11 \text{ sec.}$$

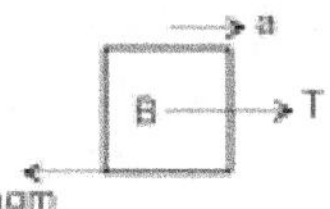


**Case-I**

$T - \mu mg = ma$ ................... (1)

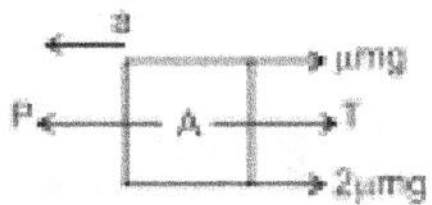


$P - T - 3\,\mu mg = ma$
puting value of T from (1)
$P - ma - \mu mg - 3\mu mg = ma$
$P - 4\,\mu mg$
$a = -2\mu g$ ................... (2)

**Case-II**

Rest

B

P
A
μmg
μ(2m)g

$a = \dfrac{P - 3\mu mg}{m}$ ................... (3)

According to Q.
accelaration is same in both cases
Hence equating the equation (2) & (3)
$P = 2\mu mg$

3.

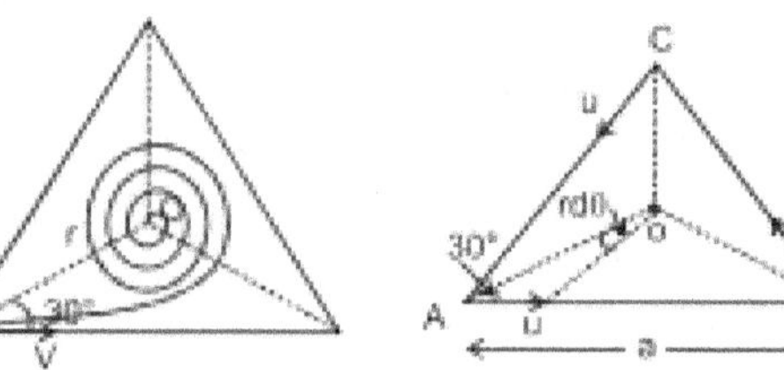

$AB = 1/2\, g(T/2)^2 = 1/8\, gT^2$

$CD = 1/2\, gT^2$

$CD/AB = 4$

4.

$F\cos\alpha - \mu N - mg\sin\theta = 0 \qquad \text{...... (i)}$

$\&\ \ N + F\sin\alpha - mg\cos\theta = 0 \qquad \text{...... (ii)}$

Solving (i) & (ii)

$$F = \frac{mg\sin\theta + \mu mg\cos\theta}{\cos\alpha + \sin\alpha}$$

$$F_{min} = \frac{mg\sin\theta + \mu mg\cos\theta}{\sqrt{1+\mu^2}} \qquad \textbf{Ans}$$

$\&\ \ \tan\alpha = \mu \qquad \Rightarrow \alpha = \tan^{-1}\mu \qquad \textbf{Ans}$

5. Let consider B as observer

$d_{min} = 10\sin 60\ \text{km} = 5\sqrt{3}$

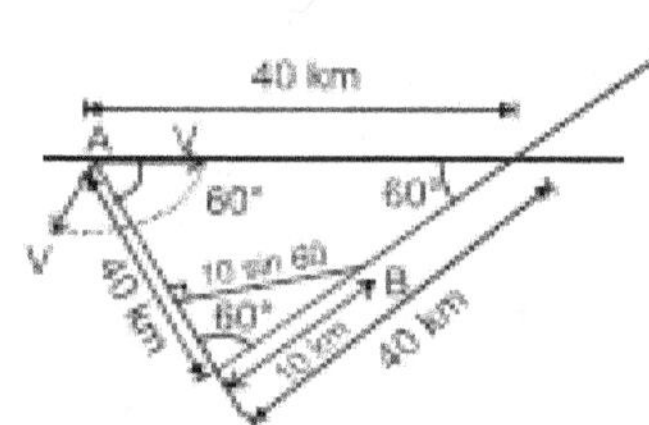

6.

$$\frac{dr}{dt} = -v\cos 30^\circ = -\frac{\sqrt{3}}{2}v$$

$$r\frac{d\theta}{dt} = v\sin 30^\circ = v/2$$

$$\frac{1}{r}\frac{dr}{d\theta} = -\sqrt{3}$$

$$\int_{r_0}^{r}\frac{dr}{r} = -\sqrt{3}\int_{0}^{\theta}d\theta \quad \Rightarrow \quad r = r_0\, e^{-\sqrt{3}\theta}$$

When A completes one revolution $\theta = 2\pi$

Time taken $t = \dfrac{r_0(1-e^{-2\sqrt{3}\pi})}{\sqrt{3}v/2}$

Distance travelled $D = vt = \dfrac{2r_0}{\sqrt{3}}(1-e^{-2\sqrt{3}\pi})$

$$D = \frac{2a}{3}(1-e^{-2\sqrt{3}\pi})$$

equation $y = x\tan\theta\left(1 - \dfrac{x}{R}\right)$

at B   $x = y$

$\tan\theta = \dfrac{R}{R - y}$      ...... (i)

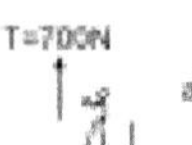


$\tan 45° = \dfrac{y}{x}$

$x = y$      ...... (ii)

$\left(\dfrac{1}{3}\right) = \dfrac{y}{R - x}$      ...... (iii)

Solving equation 2 and 3

$R = 4y = 4x$     Put in (i)

$\tan\theta = \dfrac{R}{R - \dfrac{R}{4}}$

$\tan\theta = \dfrac{4}{3}$

$\theta = 53°$

For student A to just lift off the floor, tension T in string must be greater than or equal to 700 N.
The F.B.D. of student B is
Applying Newton's second law

       $T - mg = ma$     $\Rightarrow$    $700 - 600 = 60\,a$

or      $a = \dfrac{5}{3}$ m/s²

T=700N

mg=600N

The magnitude of the force (from the string) is $T = 30$N.
The x-component $= T\sin\theta = 30 \times 3/5 = 18$N.
The y-component $= T\cos\theta = 30 \times 4/5 = 24$N.
The total force on the block is:
the x-component $= 18$N.
the y-component $= 24 - mg = 24 - 20 = 4$N.
The x-component of the acceleration $= 18\text{N}/2\text{kg} = 9\text{m/s}^2$.
The y-component of the acceleration $= 4\text{N}/2\text{kg} = 2\text{m/s}^2$.

**0.**    If stone always moves away from thrower then

$\Rightarrow \dfrac{d|\vec{r}|}{dt} > 0$

$\Rightarrow \vec{r}.\vec{v} > 0$    $\vec{r} = u\cos\theta\, t\,\hat{i} + \left(u\sin\theta\, t - \dfrac{1}{2}gt^2\right)\hat{j}$

$\vec{v} = u\cos\theta\,\hat{i} + (u\sin\theta - gt)\hat{j}$

$\vec{r}.\vec{v} = u^2 t - \dfrac{3}{2}\,ug\sin\theta\, t^2 + \dfrac{g^2}{2}t^3 > 0$

$\Rightarrow \dfrac{g^2}{2}t^2 - \dfrac{3}{2}\,ug\sin\theta\, t + u^2 > 0$

$\sin^2\theta < \dfrac{8}{9} \Rightarrow \theta < \sin^{-1}\left(\dfrac{2\sqrt{2}}{3}\right)$

11. Let total distance travelled is 4s.

$$2s \to V_1 \to t_1 = \frac{2s}{V_1}$$

$$s \to V_2 \to t_2 = \frac{s}{V_2}$$

$$s\begin{cases} V_1 \to t_0 \\ V_2 \to t_0 \end{cases} \qquad (V_1 + V_2)\, t_0 = s \Rightarrow t_0 = \frac{s}{V_1 + V_2}$$

$$<V> = \frac{4s}{t_1 + t_2 + 2t_0} = \frac{4s}{\dfrac{2s}{V_1} + \dfrac{s}{V_2} + \dfrac{2s}{V_1 + V_2}}$$

$$= \frac{4V_1 V_2 (V_1 + V_2)}{2V_2(V_1 + V_2) + V_1(V_1 + V_2) + 2V_1 V_2}$$

$$= \frac{4V_1 V_2 (V_1 + V_2)}{2V_1 V_2 + 2V_2^2 + V_1^2 + V_1 V_2 + 2V_1 V_2} = \frac{4V_1 V_2 (V_1 + V_2)}{V_1^2 + 2V_2^2 + 5V_1 V_2}$$

12. $f_s$ for 8 kg $= 0.5 \times 8(10 + 2) = 48$ N
$f_s$ for 6 kg $= 0.6 \times 6\,(10 + 2) = 43.2$ N
It can be verified that limiting friction will act on 6 kg
From FBD, tension $= 16.8$ N

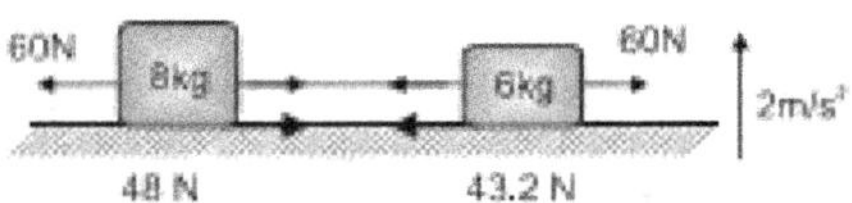


13.

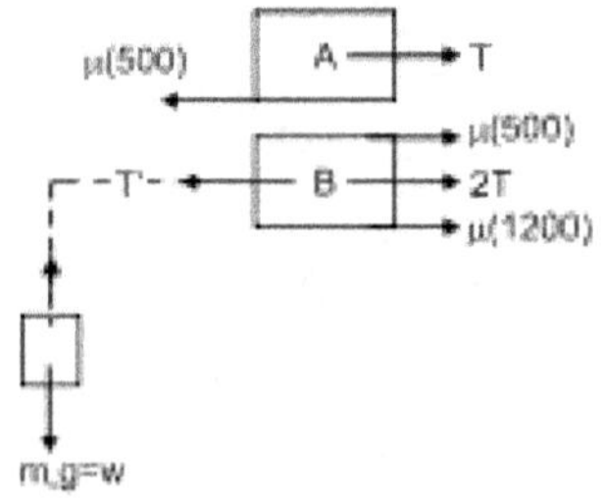


$3T + 0.3 \times 1200 = m_0 g = W$ and $T = \mu(500) = 0.3 \times 500$
$W = m_0 g = 810$ N.

14. For motion between AB

$$ma = mg \sin\alpha - \frac{\tan\alpha}{2}\, mg \cos\alpha$$

$$a = \frac{g\sin\alpha}{2} \text{ (downward)}$$

For motion between BO

$$ma = \frac{3\tan\alpha}{2}\, mg \cos\alpha - mg \sin\alpha$$

$$a = \frac{g\sin\alpha}{2} \text{ (upward)}$$

The velocity increases from zero to maximum value at B and then starts decreasing with same rate and finally becomes zero at O.

$$mv\frac{dv}{dx} = ma - kx$$

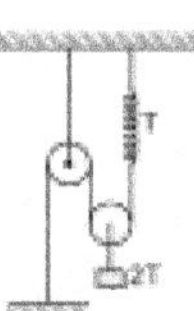


$$\int_{0}^{a} mvdv = \int_{0}^{x}(ma - kx)dx$$

$$x = \frac{2ma}{k}.$$

Initially the block is at rest under action of force 2T upward and mg downwards. When the block is pulled downwards by x, the spring extends by 2x. Hence tension T increases by 2kx. Thus the net unbalanced force on block of mass m is 4kx.

$$\therefore \qquad \text{acceleration of the block is} = \frac{4kx}{m}$$

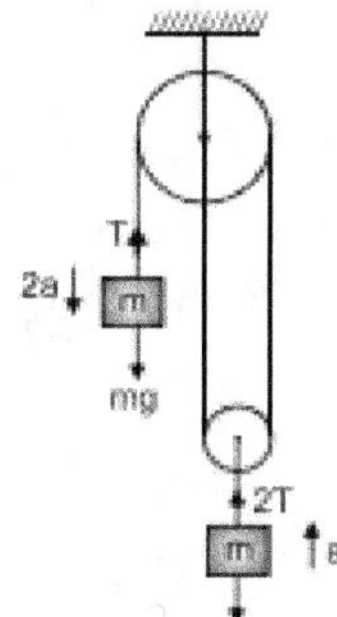


$$mg - T = 2ma \qquad \ldots\ldots(i)$$
$$2T - mg = ma \qquad \ldots\ldots(ii)$$

Solving,
$$mg = 5\,ma$$

$$a = \frac{g}{5}$$

$$T = mg - 2ma$$

$$= mg - 2m\frac{g}{5} = \frac{3mg}{5}.$$



(i)  Relative initial velocity = 5 m/s, relative final velocity = 0
Relative displacement = 50 m
Relative acceleration = constant

$$\Rightarrow \quad 50 = \left(\frac{5+0}{2}\right)t \qquad \Rightarrow \quad t = 20 \text{ sec.} \qquad \textbf{Ans.}$$

(ii)  Distance of dead line from car $C_1 = \left(\frac{25+0}{2}\right) \times 20 = 250 \text{ m.} \qquad \textbf{Ans.}$

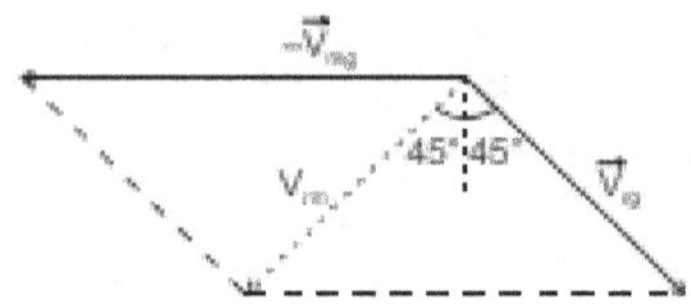



**20.**

$$\vec{V}_{rg} = \vec{V}_{rm} + \vec{V}_{mg}$$

$$\vec{V}_{rm} = \vec{V}_{rg} - \vec{V}_{mg}$$

$$V_{rm}\cos 45° = V_{rg}\cos 45°$$

$$V_{rm} = 2\sqrt{2}\ \text{m/s} = V_{rg}$$

$$V_{rm}\cos 45° = V_{mg} - V_{rg}\cos 45°$$

$$V_{mg} = 2\sqrt{2}\,\frac{1}{\sqrt{2}} + 2\sqrt{2}\,\frac{1}{\sqrt{2}} = 4\ \text{m/s}$$

using $v^2 = u^2 + 2as$ for the motion of man,

$s = 16$ m.

**21.** Let a be acceleration of system and T be tension in, the string.
F.B.D of block A

$$mg\sin 30° + T = ma$$

$$\frac{mg}{2} + T = ma \qquad ..... \text{(i)}$$

F.B.D of block B

$$mg - T = ma \qquad ..... \text{(ii)}$$

Adding equation (i) & (ii); we get

$$2ma = \frac{3mg}{2} \quad \Rightarrow \quad a = \frac{3}{4}g$$

from equation (i);

$$T = \frac{mg}{4}$$

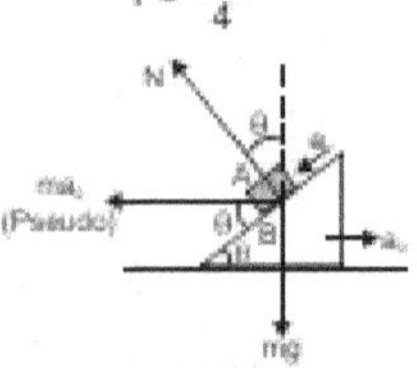

**22.**

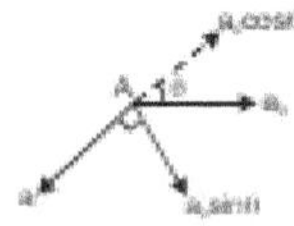

$$ma_0\sin\theta + N = mg\cos\theta \qquad \Rightarrow \qquad N = mg\cos\theta - ma_0\sin\theta$$
$$\Rightarrow \qquad N < mg\cos\theta$$

Hence, (D) is true.

$$ma_0\cos\theta + mg\sin\theta = ma$$
$$\Rightarrow \qquad a = g\sin\theta + a_0\cos\theta$$

Hence acceleration of A

$$= \sqrt{(a - a_0\cos\theta)^2 + (a_0\sin\theta)^2} > g\sin\theta .$$

**3.**

$$T = \frac{mg}{2}$$

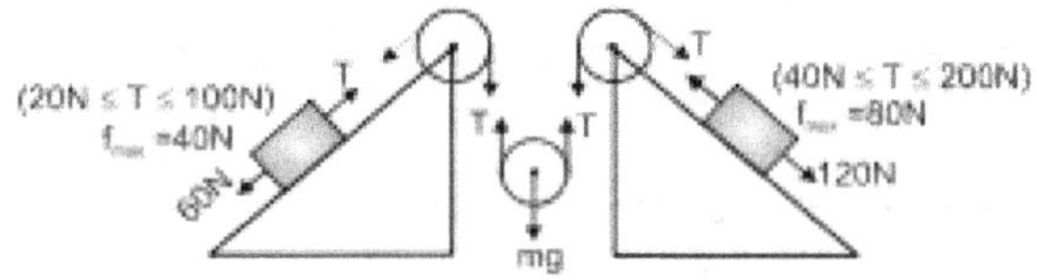

For the equilibrium of 10kg block tension in string should be between 20 N to 100 N, while for the equilibrium of 20 kg range of tension is 40 N to 200 N, so for the equilibrium of system, tension in the string must be between 40 N to 100 N and mass of block must be between 8 kg to 20 kg.

**4.**

$$20g\sin\theta + f_3 = T$$
$$20g\sin\theta + \mu(20g\cos\theta) = T$$
$$80g\sin\theta = \mu(100g\cos\theta) + \mu(20g\cos\theta)$$

$$\tan\theta = \frac{3}{8}$$

$$T = 20g\sin\theta + \mu\, 20g\cos\theta$$

$$= 20g\sin\theta + \frac{1}{4} \times 20 \times g \times \frac{8}{3}\sin\theta$$

$$= \left(\frac{100}{3}g\sin\theta\right)N$$

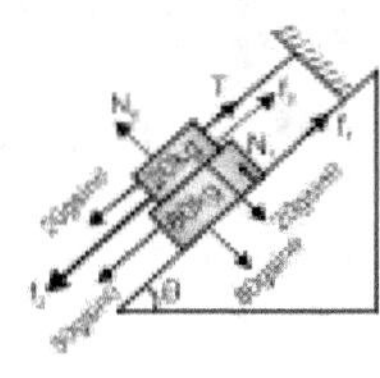


Net friction on 80 kg $= f_1 + f_3 = 80\, g\sin\theta$

force on 80 kg due to 20 kg is $\sqrt{(20g\cos\theta)^2 + (\mu 20g\sin\theta)^2}$ ...

**5.**

$$\text{Impulse} = \int \vec{F}dt = m(\vec{v}_f - \vec{v}_i)$$

$$-mg \times \text{Area under } \mu - t \text{ graph} = m\,(v_f - 20.5)$$

$$-mg \times \left[\frac{1}{2}(0.4 + 0.3) \times 1 + 0.4 \times 2 + \frac{1}{2}(0.4 + 0.2) \times 1\right] = m(v_f - 20.5)$$

$$v_f = 6m/s$$

**6.**

$$x = t^3/3 - 3t^2 + 8t + 4$$
$$v = t^2 - 6t + 8 = (t-2)(t-4)$$
$$a = 2(t-3)$$

|   |   | 2 |   | 3 |   | 4 |   |
|---|---|---|---|---|---|---|---|
| V | + |   | − |   | − |   | + |
| a | − |   | − |   | + |   | + |

$$S_1 = \left(\frac{32}{3} - 4\right) + \left(\frac{32}{3} - \frac{28}{3}\right) + \left(\frac{32}{3} - \frac{28}{3}\right) = \frac{20}{3} + \frac{8}{3} = \frac{28}{3}\ \text{m.}$$

$$S_2 = \left(\frac{32}{3} - 4\right) + \left(10 - \frac{28}{3}\right) = \frac{20}{3} + \frac{2}{3} = \frac{22}{3}\ \text{m}$$

$$\frac{S_1}{S_2} = \frac{28}{22} = \frac{14}{11} = \frac{3\alpha + 2}{11} \Rightarrow \alpha = 4$$

27. The block begins to slide if
$$F \cos 37° = \mu \,(mg - F \sin 37°)$$
$$5t \,[\cos 37° + \mu \sin 37°] = \mu \, mg$$

$$5t \left[\frac{4}{5} + \frac{3}{5}\right] = 70 \qquad \text{or} \qquad t = 10 \text{ second}$$

28. Taking block + wedge as system and applying NLM in horizontal direction
$$f_2 = m_1 a \cos\theta$$
$$= m_1 \,[g(\sin\theta - \mu_1 \cos\theta)] \cos\theta \qquad \dots\dots (1)$$
Again applying NLM in vertical direction
$$(m_1 + m_2)g - N_2 = m_1 a \sin\theta$$
$$N_2 = (m_1 + m_2)g - m_1 \sin\theta(g\sin\theta - \mu_1 g \cos\theta)$$
For limiting condition $f_2 = \mu_2 N_2$ $\quad\dots\dots (2)$
From (1) and (2)
$$\mu_2 = \frac{m_1 \cos\theta(g\sin\theta - \mu_1 g\cos\theta)}{(m_2 + m_2)g - m_1 \sin\theta(g\sin\theta - \mu_1 g\cos\theta)}$$

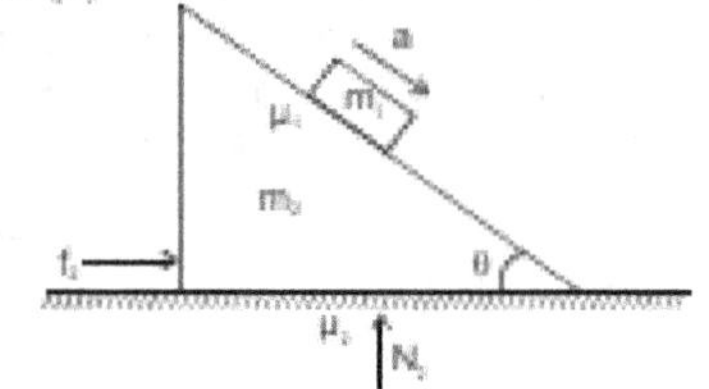

Using values
$$\mu_2 = \frac{1}{8} = 125 \times 10^{-3}$$

**Ans.** 125

29.

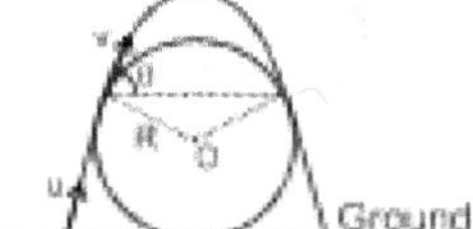


$$\frac{2v^2 \sin\theta\cos\theta}{g} = 2R\sin\theta \;\Rightarrow\; v^2 = \frac{Rg}{\cos\theta}$$
$$u^2 = v^2 + 2g R\,(1 + \cos\theta)$$
$$u^2 = \frac{Rg}{\cos\theta} + 2gR + 2gR\cos\theta$$
$$u^2 = Rg\left(\frac{1 + 2\cos^2\theta}{\cos\theta}\right) + 2gR$$

for u to be minimum $\dfrac{1 + 2\cos^2\theta}{\cos\theta} = \min$

$$\Rightarrow \quad \cos\theta = \frac{1}{\sqrt{2}} \quad \Rightarrow \quad \theta = \pi/4$$

$$u_{\min} = \sqrt{\sqrt{2}Rg + 2gR + \sqrt{2}Rg} = \sqrt{2gR(\sqrt{2} + 1)}$$

Let everything moves together
Then,

$$20N \leftarrow \boxed{12} \rightarrow 8N$$

$$a = \frac{12}{12} = 1 \text{ m/s}^2$$

$$5N \leftarrow \boxed{\begin{array}{c} \leftarrow 1\text{m/s}^2 \\ C \end{array}} \qquad f_{AB} = 17N \leftarrow \boxed{\begin{array}{c} \leftarrow 1\text{m/s}^2 \\ B \end{array}} \begin{array}{l} \rightarrow 5N \\ \rightarrow 8N \end{array}$$

But $f_{AB \text{ maximum}} = 15N$
So, sliding occurs.
Now, see if B and C move together.

$$15N \leftarrow \boxed{\begin{array}{c} B \\ \hline C \end{array}} \rightarrow 8N$$

$$a = \frac{15 - 8}{9} = \frac{7}{9} \text{ m/s}^2$$

So, friction acting between B and C is $\frac{7}{9} \times 5$ m/s$^2$.

$a = b + c$

Net acceleration of $A = \sqrt{a^2 + c^2 + 2ac \cos(\pi - \theta)} = \sqrt{(b + c)^2 + c^2 - 2(b + c).c.\cos\theta.} = \sqrt{3}$

For block B. ;

$$2ma_B = F - \frac{mg}{2}$$

$$a_B = g$$

For block A ;

$$ma_A = mg$$
$$a_A = g/2$$
$$a_{AB} = -g/2$$

$$L = \frac{1}{2}\frac{g}{2}t_1^2$$

$$t_1 = \sqrt{\frac{2}{5}} \text{ s}$$

time of flight $t_2 = \sqrt{\frac{2h}{g}} = \frac{1}{\sqrt{10}}$ s

Velocity when A leaves B. ;

$$V_A = g/2 \, t_1 = g/2 \times \sqrt{\frac{4L}{g}} = \sqrt{10} \text{ m/s}$$

$$S_x = V_A t_2 = 1 \text{m}$$

$$S_y = \frac{1}{2}\frac{g}{2}\frac{2h}{g} = \frac{1}{4}\text{m}$$

$$\frac{S_x}{S_y} = 4$$

$$\vec{a_A} = -\frac{g}{2}\hat{j} - g\hat{k}$$

$$\vec{a_B} = -\frac{5g}{4}\hat{i}, |\vec{a_{AB}}| = \left(\sqrt{\frac{1}{4} + 1 + \frac{25}{16}}\right)g$$

**34 to 36.** If we draw FBD of block w.r.t wedge

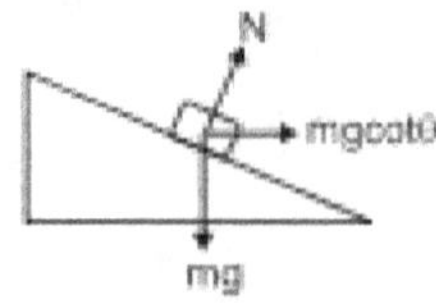

$$N + F_s \sin\theta = mg\cos\theta$$
$\Rightarrow \quad N = 0$
so   w.r.t ground block will fall freely.

$$h = \frac{1}{2}gt^2 \quad \text{and} \quad h = \ell\sin\theta$$

**37. to 39**

From conservation of momentum

$$3mv = mu \qquad\qquad \text{or} \quad v = \frac{u}{3}$$

Net workdone by friction $= \dfrac{1}{2}\,3m\left(\dfrac{u}{3}\right)^2 - \dfrac{1}{2}mu^2 = -\dfrac{1}{3}mu^2$

net work done by friction $= \displaystyle\int_{\ell}^{0} \mu(x\lambda g)(-dx) = -\mu\lambda g\,\dfrac{L^2}{2}$

Also magnitude of net work done by friction $= \mu\lambda g\,\dfrac{L^2}{2} = \mu mg\,\dfrac{L}{2}$

$$\therefore \ \frac{1}{3}mu^2 = \mu mg\,\frac{L}{2} \qquad \text{or} \quad \mu = \frac{2}{3}\frac{u^2}{gL}$$

$$3mv = mu \qquad\qquad \text{or} \quad v = \frac{u}{3}$$

**40.**  $\left|\vec{F_1} + \vec{F_2}\right| < |f_1|_{max} + |f_2|_{max}$

So, both blocks not move in any case.

$|f_1|_{max} = 50\,N \qquad ; |f_2|_{max} = 100\,N$

(A)


(B)


(C)


(D)

**41.** $\vec{V}_{p,p} = V_2\hat{i} + 25\hat{j} + V_1\hat{k}$

$\vec{a}_{p,p} = -2\hat{i} - 12.5\hat{j}$

$\vec{V}_{p,p}$ = Velocity of particle relative to platform

$$\text{Time} = \frac{2 \times 25}{12.5} = 4 \text{ sec.}$$

$$8 \leq V_2 \times 4 - \frac{1}{2} \times 2 \times 4^2 \leq 16$$

$$6 \leq V_2 \leq 8$$
$$16 \leq V_1 \times 4 \leq 24$$
$$4 \leq V_1 \leq 6$$

$$Y = 25 \times 4 - \frac{1}{2} \times 10 \times 4^2 = 100 - 80 = 20m$$

**42.** (A) Q (b) Q (C) R (D) S
FBD's

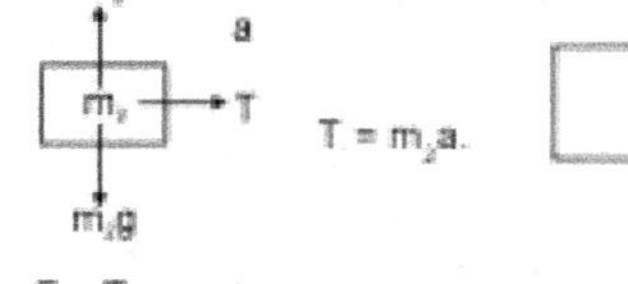


$F - T = m_1a$
$F = (m_1 + m_2)a$
$\therefore \quad T = m_2a$

$$\Rightarrow \quad a = \frac{F}{m_1 + m_2}$$

$$\therefore T = \frac{m_2F}{m_1 + m_2}.$$

$$F_s = 0, \ a_M = 0$$

# Mock Test 3

1. A solid object is rotating freely without experiencing any external torque. In this case
   (A) both the angular momentum and angular velocity have constant direction.
   (B) the direction of the angular momentum is constant but the direction of the angular velocity might not be constant.
   (C) the direction of the angular velocity is constant but the direction of the angular momentum might not be constant.
   (D) neither the angular momentum nor the angular velocity necessarily has constant direction.

2. A 10-cm cube of metal is fastened rigidly in place. A second, identical cube of metal is pulled across the top of the first cube at a constant speed by a constant 10 N force, as shown. The frictional force between the cubes
   (A) is less than 10 N.
   (B) is equal to 10 N.
   (C) is greater than 10 N.
   (D) cannot be determined without a detailed model of the two surfaces.

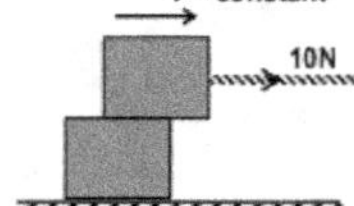

3. In a lake, stream direction is as shown in the figure. A man starting from the point P on the bank 1 wants to move to the bank 2 in shortest time. He should swim
   (A) along PQ
   (B) along PR
   (C) In a direction in-between PQ and PR
   (D) in all cases he would reach at the same time

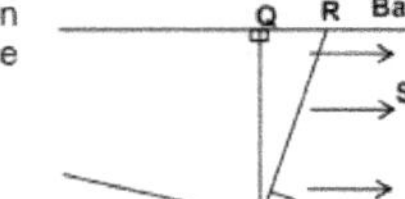

4. A particle is projected vertically upwards from the ground with a speed v and a second particle is projected at the same instant from a height h directly above the first particle with the same speed v at an angle of projection $\theta$ with the horizontal in upward direction. The time when the distance between them is minimum is

   (A) $\dfrac{h}{2v\sin\theta}$
   (B) $\dfrac{h}{2v\cos\theta}$
   (C) $\dfrac{h}{\sqrt{2}\,v}$
   (D) $\dfrac{h}{2v}$

5. Which of the following forces is non conservative one?

   (A) $3\hat{i} + 4\hat{j}$
   (B) $4x\hat{i} + 3y\hat{i}$
   (C) $3x^2\,\hat{i} + 4y^2\hat{j}$
   (D) $y^2\hat{i} + x^2\hat{j}$

6. In the figure shown, the wedge is fixed and the masses are released from rest. The coefficient of friction between A and the incline is $\mu_1$ and B and the incline is $\mu_2$. Then which of the following sentences is (are) correct
   I. normal reaction between A and B can never be zero
   II. normal reaction between A and B is zero only if $\mu_1 = \mu_2$
   (A) I only
   (B) II only
   (C) both I and II
   (D) none

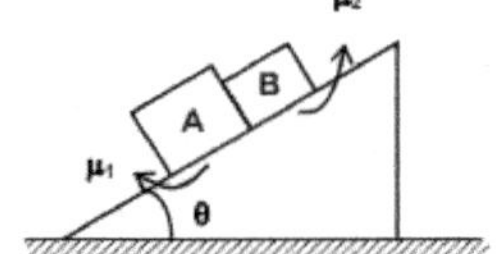

7. A disk is fixed at its centre O and rotating with constant angular velocity $\omega$. There is a rod whose one end is connected at A on the disc and the other end is connected with a ring which can freely move along the fixed vertical smooth rod. At an instant when the rod is making an angle 30° with the vertical the ring is found to have a velocity v in the upward direction. Find $\omega$ of the disk. Given that the point A is R/2 distance above point O and length of the rod AB is $\ell$

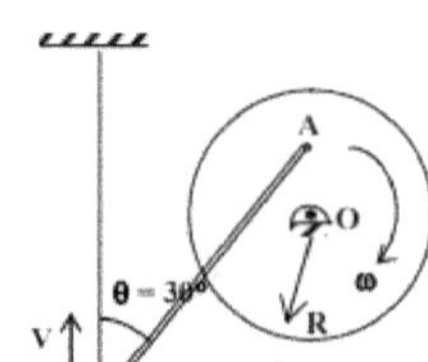

   (A) $\dfrac{v\sqrt{3}}{2\ell}$
   (B) $\dfrac{v\sqrt{3}}{2R}$
   (C) $\dfrac{2\sqrt{3}v}{R}$
   (D) $\dfrac{2v}{\ell\sqrt{3}}$

Magnetic field B in a cylindrical region of radius r varies according to the law $B = B_0 t$ as shown in the figure. A fixed conducting loop ABCDA of resistance R is lying in the region as shown. The current flowing through the loop is

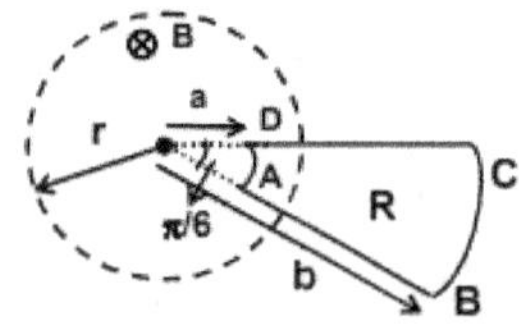

(A) $\dfrac{\pi a^2 B_0}{12R}$

(B) $\dfrac{\pi (r^2 - a^2) B_0}{12R}$

(C) $\dfrac{\pi (b^2 - a^2) B_0}{12R}$

(D) none of the above

In the circuit shown in the adjacent figure, the batteries are ideal. The charge on the capacitor C is

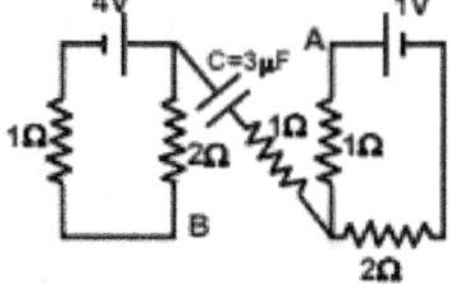

(A) 10 μC

(B) 20 μC

(C) 30 μC

(D) zero

A parallel plate capacitor of area A and separation d is provided with thin insulating spacers to keep its plates aligned in an environment of fluctuating temperature. If the coefficient of thermal expansion of material of the plate is $\alpha$, find the coefficient of thermal expansion ($\alpha_s$) of the spacers in order that the capacitance does not vary with temperature. (Ignore effect of the spacers on capacitance.)

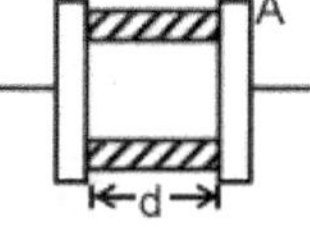

(A) $\alpha_s = \dfrac{\alpha}{2}$

(B) $\alpha_s = 3\alpha$

(C) $\alpha_s = 2\alpha$

(D) $\alpha_s = \alpha$

A T shaped object with dimensions shown in the figure is lying on a smooth floor. A force $\vec{F}$ is applied at the point P parallel to AB such that the object has only the translational motion without rotation. Find the location of P with respect to C.

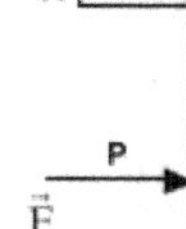

(A) $\dfrac{3}{4}\ell$

(B) $\ell$

(C) $\dfrac{4}{3}\ell$

(D) $\dfrac{3}{2}\ell$

12. An infinite thread of charge density $\lambda$ lies along z-axis. The potential difference between points A (4, 3, 4) and B (3, 4, 0) is

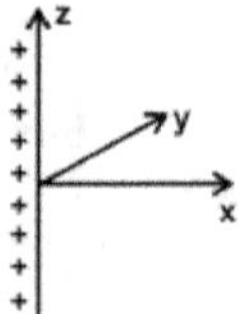

(A) $\dfrac{\lambda}{2\pi \epsilon_0} \ln\left(\dfrac{\sqrt{41}}{5}\right)$

(B) $\dfrac{\lambda}{2\pi \epsilon_0} \ln(5)$

(C) zero

(D) $\dfrac{\lambda}{2\pi \epsilon_0}$

13. The sun having surface temperature $T_S$ radiates like a black body. The radius of sun is $R_S$ and earth is at a distance R from the surface of sun. Earth absorbs radiations falling on its surface from sun only and is at constant temperature T. If radiations falling on earth's surface are almost parallel and earth also radiates like a blackbody, then

(A) $T = T_s\sqrt{\dfrac{R_s}{2R}}$

(B) $T = T_s$

(C) $T = \dfrac{T_s}{2}\sqrt{\dfrac{R_s}{R}}$

(D) $T = T_s\sqrt{\dfrac{R_s}{R}}$

14. In the adjacent figure, the mutual inductance of the infinite straight wire and the coil is M, while the self inductance of the coil is L. The current in infinite wire is varying according to the relation $I_1 = \alpha t$, where $\alpha$ is a constant and t is the time. The time dependence of current in the coil is

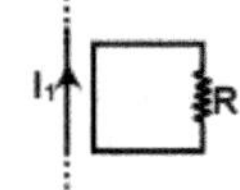

(A) $\dfrac{M\alpha}{R}$

(B) $\dfrac{M\alpha L}{R} e^{-Rt/L}$

(C) $\dfrac{\alpha}{R}\left(1 - e^{-Rt/L}\right)$

(D) none of the above

15. A cannon-ball is fired with a speed 50 m/sec with respect to ground from one end of a stationary railroad car which rests on a smooth track. The ball collides with the opposite wall and stops. How much distance did the railroad car cover just before the collision? Given that the ratio of the mass of the ball to that of the railroad car is 1/99 and the length of the car is 50 m

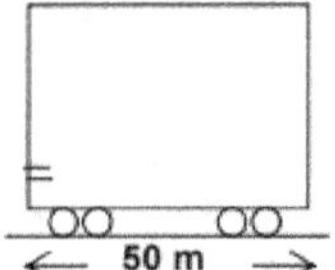

(A) 2 m

(B) $\dfrac{1}{2}$ m

(C) 0 m

(D) cannot be calculated from acts

A cubical box of side 1 m contains an ideal gas at pressure 100 N/m². If $\Sigma v_x^2 = \Sigma v_y^2 = \Sigma v_z^2 = 10^{28}\ m^2/s^2$, where $v_x$, $v_y$ and $v_z$ are the x, y and z components, respectively, of the gas molecule, then the mass of each gas molecule is

(A) $10^{-20}$ g

(B) $10^{-23}$ g

(C) $10^{-18}$ g

(D) $10^{-26}$ g

Two fixed insulating rings A and B carry charges with uniform linear charge density $+\lambda$ and $-\lambda$, respectively, as shown in the adjacent figure. The planes of the rings are parallel to each other and their axes are coinciding. A particle of charge "q" and mass "m" is released with zero velocity from centre P of the positively charged ring. The kinetic energy of the particle when it reaches centre Q of the negatively charged ring will be

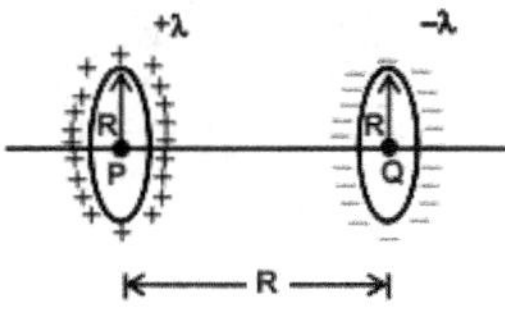


(A) $\sqrt{\dfrac{2\lambda q}{m\,\epsilon_0}\left(1-\dfrac{1}{\sqrt{2}}\right)}$

(B) $\dfrac{\lambda}{\epsilon_0}\left(1-\dfrac{1}{\sqrt{2}}\right)q$

(C) $\dfrac{\lambda}{2\epsilon_0}\left(1-\dfrac{1}{\sqrt{2}}\right)q$

(D) none of the above

Magnetic field in the cylindrical region with its axis passing through O varies at a constant rate $\dfrac{dB}{dt}$. A triangular imaginary loop ABC, with AB = BC, is lying in this region as shown in the adjacent figure. The work done to move unit positive charge from A to B along the side AB is

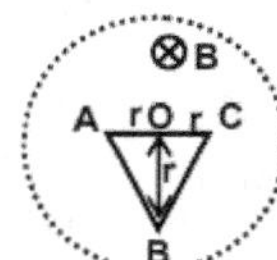


(A) $r^2\dfrac{dB}{dt}$

(B) $\pi r^2\dfrac{dB}{dt}$

(C) $\dfrac{r^2}{2}\dfrac{dB}{dt}$

(D) $\dfrac{3r^2}{4}\dfrac{dB}{dt}$

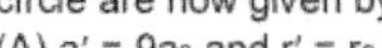

A puck is moving in a circle of radius $r_0$ with a constant speed $v_0$ on a level frictionless table. A string is attached to the puck, which holds it in the circle; the string passes through a frictionless hole and is attached on the other end to a hanging object of mass M. The puck is now made to move with a speed $v' = 3v_0$, but still in circle. The mass of the hanging object is left unchanged. The acceleration a' of the puck and the radius r' of the circle are now given by

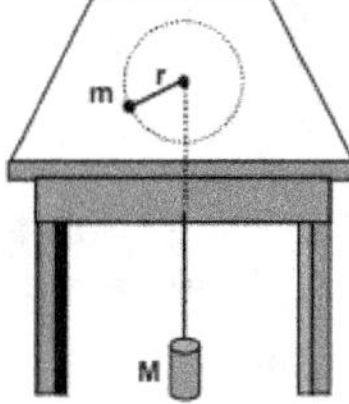


(A) $a' = 9a_0$ and $r' = r_0$.

(B) $a' = a_0$ and $r' = 9r_0$

(C) $a' = 9a_0$ and $r' = 9r_0$

(D) $a' = a_0$ and $r' = r_0$

20. Two spherical black bodies A and B are emitting radiations at same rate. The radius of B is doubled keeping radius of A fixed. The wavelength corresponding to maximum intensity becomes half for A while it remains same for B. Then, the ratio of rate of radiation energy emitted by A and B is

(A) 2                                        (B) 1/2
(C) 4                                        (D) 1/4

21. A charge +q is placed at a distance 'd' from the centre of the uncharged metallic cube of side 'a'. The electric field at the centre of the cube due to induced charges on the cube will be

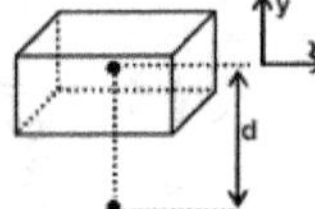

(A) zero

(B) $\dfrac{q}{4\pi \epsilon_0 d^2}\left(-\hat{j}\right)$

(C) $\dfrac{q}{4\pi \epsilon_0 d^2}\left(\hat{j}\right)$

(D) $\dfrac{q}{4\pi \epsilon_0 \left(d-\dfrac{a}{2}\right)^2}\left(-\hat{j}\right)$

22. A physical quantity $\rho$ is calculated by using the formula $\rho = \dfrac{1}{10}\dfrac{xy^2}{z^{1/3}}$, where x, y and z are experimentally measured quantities. If the fractional error in the measurement of x, y and z are 2%, 1% and 3%, respectively, then the maximum fractional error in the calculation of $\rho$ is

(A) 0.5%                                  (B) 5%
(C) 6%                                    (D) 7%

23. A uniform square plate of mass m is supported in a horizontal plane by a vertical pin at B and is attached at A to a spring of constant K. If corner A is given a small displacement and released, determine the period of the resulting motion.

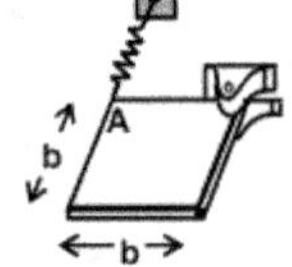

(A) $2\pi\sqrt{\dfrac{2Mb}{3K}}$

(B) $2\pi\sqrt{\dfrac{Mb}{6K}}$

(C) $\pi\sqrt{\dfrac{Mb}{3K}}$

(D) none of the above

The adjacent figure shows cross section of a hollow glass tube of 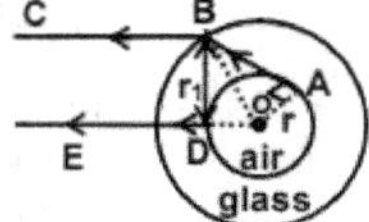 internal radius r, external radius R and index of refraction n. For two rays DE and ABC (in which DE lies on ODE and DE and BC are parallel), the separation $r_1$ will be

(A) $r_1 = (n - 1)R$ 　　　　　　　　　　　　(B) $r_1 = n^2R$
(C) $r_1 = nr$ 　　　　　　　　　　　　　　　(D) $r_1 = n^2r$

The given figure shows several possible elliptical orbits of a satellite. On which orbit will the satellite acquire the largest speed? 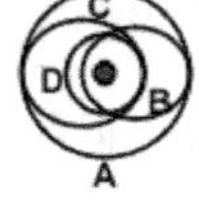

(A) A 　　　　　　　　　　　　　　　　　(B) B
(C) C 　　　　　　　　　　　　　　　　　(D) D

An object is projected from A and reaches B on the same horizontal surface. A plane mirror inclined at an angle $\theta$ to the horizontal is placed on the ground as shown in the figure. Which of the following best describes the path of the image? 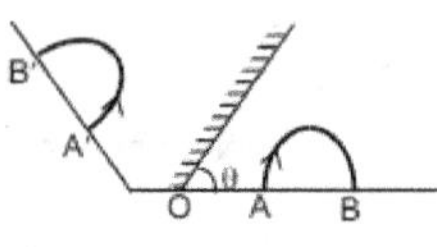

(A) 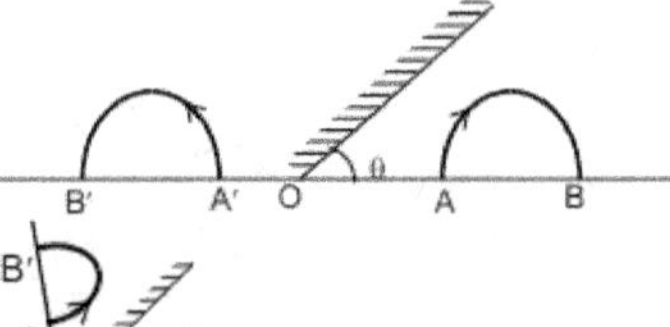　　　　　　(B) 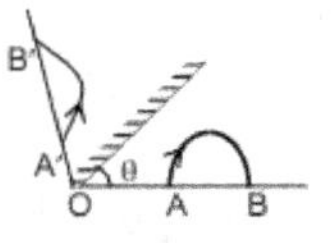

(C) 　　　　　　　　　　　　　　　　(D)

A solid uniform ball of volume V floats on the interface of two immiscible liquids [The specific gravity of the upper liquid is $\gamma/2$ and that of the lower liquid is $2\gamma$, where $\gamma$ is the specific gravity of the solid ball.] The fraction of the volume of the ball that will be in upper liquid is

(A) $\dfrac{2}{3}V$ 　　　　　　　　　　　　(B) $\dfrac{1}{3}V$

(C) $\dfrac{V}{4}$ 　　　　　　　　　　　　　(D) $\dfrac{3V}{4}$

28. In the figure shown, water flows into the open vessel through pipe A at a constant rate. The water can flow out of the vessel by pipe B. The maximum flow-rate of the water in pipe B is slightly more than that in pipe A. Final height of the water level in the vessel would be

(A) $h_0$

(B) $h_f$

(C) somewhere in-between $h_0$ and $h_f$

(D) The water level will keep oscillating in the vessel

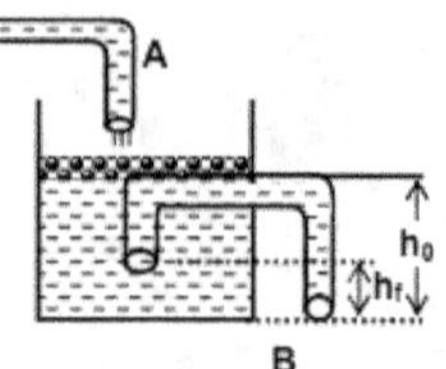


29. The earth is moving on an elliptical path, whose one focus sun is situated as shown in figure. If $AS = r_{min}$ and $SB = r_{max}$, the sun and earth system obey the Kepler's law, the square of time-period is directly proportional to

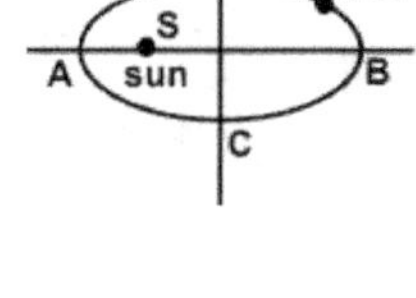


(A) $\left(r_{min}\right)^3$

(B) $\left(r_{max}\right)^3$

(C) $\left(\dfrac{r_{min} + r_{max}}{2}\right)^3$

(D) $\left(\dfrac{r_{min}\,r_{max}}{r_{min} + r_{max}}\right)^3$

30. A body is moving under the action of central force $\vec{F}(r)\hat{e}_r$ such that its position vector is $\vec{r} = r\hat{e}_r$. Then, choose the correct statement (symbols are having usual meaning and $\hat{e}_r$, $\hat{e}_\theta$ denote unit vectors along the radial and tangential direction, respectively) from the following.

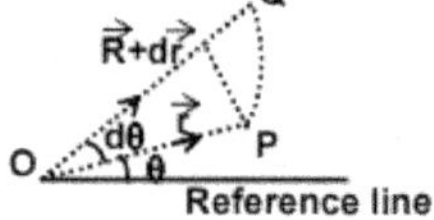


(A) $\vec{v} = \dfrac{dr}{dt}\hat{e}_r + r\dfrac{d\theta}{dt}\hat{e}_\theta$, $\quad \vec{a} = \left[\dfrac{d^2r}{dt^2} - r\left(\dfrac{d\theta}{dt}\right)^2\right]\hat{e}_r$, $\quad 2\dfrac{dr}{dt}\dfrac{d\theta}{dt} + r\dfrac{d^2\theta}{dt^2} = 0$

(B) $\vec{v} = \dfrac{dr}{dt}\hat{e}_r + r\dfrac{d\theta}{dt}\hat{e}_\theta$, $\quad \vec{a} = \left[\dfrac{d^2r}{dt^2} + r\left(\dfrac{d\theta}{dt}\right)^2\right]\hat{e}_r$, $\quad 2\dfrac{dr}{dt}\dfrac{d\theta}{dt} - r\dfrac{d^2\theta}{dt^2} = 0$

(C) $\vec{v} = \dfrac{dr}{dt}\hat{e}_r - r\dfrac{d\theta}{dt}\hat{e}_\theta$, $\quad \vec{a} = \left[\dfrac{d^2r}{dt^2} + r\left(\dfrac{d\theta}{dt}\right)^2\right]\hat{e}_r$, $\quad 2\dfrac{dr}{dt}\dfrac{d\theta}{dt} + r\dfrac{d^2\theta}{dt^2} = 0$

(D) $\vec{v} = \dfrac{dr}{dt}\hat{e}_r - r\dfrac{d\theta}{dt}\hat{e}_\theta$, $\quad \vec{a} = \left[\dfrac{d^2r}{dt^2} - r\left(\dfrac{d\theta}{dt}\right)^2\right]\hat{e}_r$, $\quad 2\dfrac{dr}{dt}\dfrac{d\theta}{dt} + r\dfrac{d^2\theta}{dt^2} = 0$

# Solution

**A**

$\tau = 0 \Rightarrow \alpha = 0 \Rightarrow \vec{\omega} = $ constant & $\vec{L} = I\vec{\omega} = $ constant

**B**

$\vec{v} = $ constant $\therefore \vec{a} = 0 \therefore \sum \vec{F} = 10 - f = 0 \Rightarrow f = 10N$

**A**

The man must swim along the direction of shortest distance.

**D**

In the relative frame

$$\vec{S}_0 = -h\hat{j}, \quad \vec{u}_{rel} = v\cos\theta\,\hat{i} + (v\sin\theta - v)\hat{j}; \quad \vec{a}_{rel} = 0$$

**D**

for conservative force $F_x = \dfrac{-dU}{dx}; F_y = \dfrac{-dU}{dy}$

$$\frac{\delta F_x}{\delta y} = \frac{\delta F_y}{\delta x} = \frac{\delta^2 U}{\delta x \delta y}$$

**D**

if $\mu_1 < \mu_2$, then also $N > 0$

**C**

**B**

$$e = \frac{d\phi}{dt} = \frac{(r^2 - a^2)}{2}\frac{\pi}{6}\frac{dB}{dt}$$

$$\therefore i = \frac{(r^2 - a^2)\pi B_0}{12R}$$

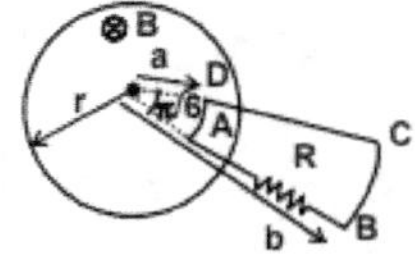

**D**

No current will flow through the capacitor, so charge will be zero.

**C**

$C = \dfrac{\epsilon_0 A}{x}$, where x is separation between plates.

$$\frac{1}{C}\frac{dC}{dT} = \frac{1}{A}\frac{dA}{dT} - \frac{1}{x}\frac{dx}{dT}$$

For $\dfrac{dC}{dT} = 0$, $\dfrac{1}{x}\dfrac{dx}{dT} = \dfrac{1}{A}\dfrac{dA}{dT}$

$\Rightarrow \alpha_S = 2\alpha$

**C**

For pure translatory motion of object, the force should act at the centre of mass of the object

$$y_{cm} = \frac{m \times 2\ell + 2m \times \ell}{3m} = \frac{4\ell}{3}$$

12.    C

Equipotential surfaces are coaxial cylinders, therefore (4, 3, 4) and (3, 4, 0) lie on equipotential surface.

13.    A

Energy absorbed per unit time by earth's surface, if its radius is r, is

$$\dot{Q}_{abs} = \frac{\sigma 4\pi R_s^2 T_s^4}{4\pi R^2} \times \pi r^2$$

Energy radiated per unit time by earth's surface, $\dot{Q}_{rad} = \sigma (4\pi r^2) T^4$

In equilibrium,

$$\dot{Q}_{abs} = \dot{Q}_{rad}$$

$$\Rightarrow \quad T^4 = \frac{T_s^4}{4} \frac{R_s^2}{R^2}$$

14.    D

15.    B

CM will be at rest, m (50 − x) = 99 m (x)

$$\Rightarrow x = \frac{1}{2} m$$

16.    B

$PV = m \, \Sigma v_i^2$

$$\Rightarrow m = \frac{100 \times 1}{10^{28}} = 10^{-26} \text{ kg} = 10^{-23} \text{ g}$$

17.    B

Potential at P due to rings

$$V_p = \frac{\lambda 2\pi R}{4\pi \, \epsilon_0 \, R} - \frac{\lambda 2\pi R}{4\pi \, \epsilon_0 \, \sqrt{2}R} = \frac{\lambda}{2 \, \epsilon_0} \left( 1 - \frac{1}{\sqrt{2}} \right)$$

Potential at Q due to rings is

$$V_Q = -\frac{\lambda}{2 \, \epsilon_0} \left( 1 - \frac{1}{\sqrt{2}} \right)$$

$\Delta(\text{K.E.}) + \left( V_Q - V_P \right) q = 0$

$$\Rightarrow \frac{1}{2} mv^2 = \frac{\lambda}{\epsilon_0} \left( 1 - \frac{1}{\sqrt{2}} \right) q$$

18.    C

No e.m.f. will be induced in AC because electric lines of force are perpendicular to AC and e.m.f. induced in AB = e.m.f. induced in BC.

$$\therefore \text{ e.m.f induced in AB} = \frac{1}{2} \left| \frac{d\phi}{dt} \right| = \frac{1}{2} r^2 \frac{dB}{dt}$$

19.    B

$$T = \frac{mv^2}{R}, \text{ and } T = mg$$

20.    C

Initially $\sigma \, 4\pi \, r_A^2 T_A^4 = \sigma 4\pi r_B^2 T_B^4$

$$\text{Now} \quad \frac{Q_A}{Q_B} = \frac{\sigma (4\pi)(r_A)^2 (2T_A)^4}{\sigma 4\pi (2r_B)^2 (T_B)^4} = \frac{4 r_A^2 T_A^4}{r_B^2 T_B^4} = 4$$

B

Net field at the centre of cube = 0

$\therefore \quad E_i + E_q = 0$

$\Rightarrow \quad E_i = -\dfrac{q}{4\pi\epsilon_0\, d^2}$

$\therefore \quad |E_i| = \dfrac{q}{4\pi\epsilon_0\, d^2}$

B

A

$-kbx = \dfrac{2Mb^2\alpha}{3}$

$\alpha = -\dfrac{3}{2}\dfrac{kx}{Mb}$

C

Use Snell's law: $\mu \sin\theta$ = constant

B

$mv^2/r = GMm/r^2$

C

A

$V = V_1 + V_2$ and $V\gamma = V_1\gamma_1 + V_2\gamma_2$

C

At first, the level of water will gradually rise to height $h_0$. After reaching height $h_0$, some of the water will get drained through the siphon. As soon as the entire cross section of the top of the siphon pipe is filled with water, the water level will begin to drop since, from given conditions, the flow-rate of the water flowing from pipe B is greater than from pipe A. As the water level drops, the pressure that drives the siphon decreases and as a result the velocity of water in pipe B decreases. Level will continue to sink until the flow rate of water through pipe B becomes equal to that through pipe A.

C

$T^2 \propto$ (semi-major axis)$^3$

A

Use $\dfrac{d\hat{e}_r}{dt} = \dfrac{d\theta}{dt}\hat{e}_\theta$ and $\dfrac{d\hat{e}_\theta}{dt} = -\dfrac{d\theta}{dt}\hat{e}_r$

# Mock Test 4

1. The range for a projectile that lands at the same elevation from which it is fired is given by $R = (u^2/g) \sin 2\theta$. Assume that the angle of projection = 30°. If the initial speed of projection is increased by 1%, while the angle of projection is decreased by 2% then the range changes by
   (A) -0.3%                                  (B) + 4.3%
   (C) +0.65%                                 (D) 0.85%

2. Two highways are perpendicular to each other: imagine them to be along the x-axis and the y-axis, respectively. At the instant t = 0, a police car P is at a distance d = 400 m from the intersection and moving at speed of 80 km/h towards it along the x-axis. Motorist M is at a distance of 600 m from the intersection and moving towards it at a speed of 60 km/h along the y-axis. The minimum distance between the cars is
   (A) 300 m                                  (B) 240 m.
   (C)  180m.                                 (D) 120 m

3. Two identical blocks are  attached by a massless string running over a pulley as shown in Figure. The rope initially runs over the pulley at its (the rope's) midpoint, and the surface that block 1 rests on is frictionless. Blocks 1 and 2 are initially at rest when block 2 is released with the string taut and horizontal. Assume that the initial distance from block 1 to the pulley is the same as the initial distance from block 2 to the wall.

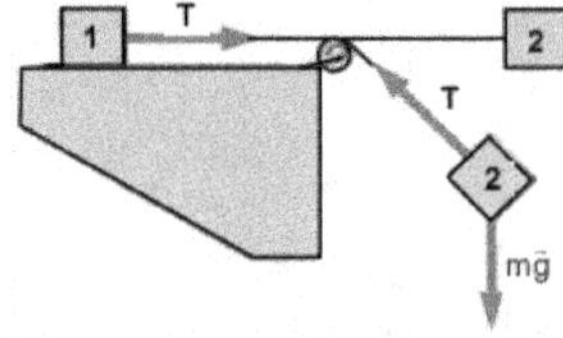


   (A) Block 1 hits the pulley before Block 2 hits the wall
   (B) Block 2 hits the wall before block 1 hits the pulley
   (C) Block 1 hits pulley simultaneously with block 2 hitting the wall
   (D) Which block hits the obstruction first depends on the actual mass of the blocks and the length of the string.

A person starts with a speed of $\sqrt{0.5gr}$ at the top of a large frictionless spherical surface, and slides into the water below (see the drawing). Then, the person
(A) loses contact when $\cos\theta = 1/3$
(B) slides through a height of r/6 before losing contact
(C) slides through a height of r/3 before losing contact
(D) never loses contact with the surface

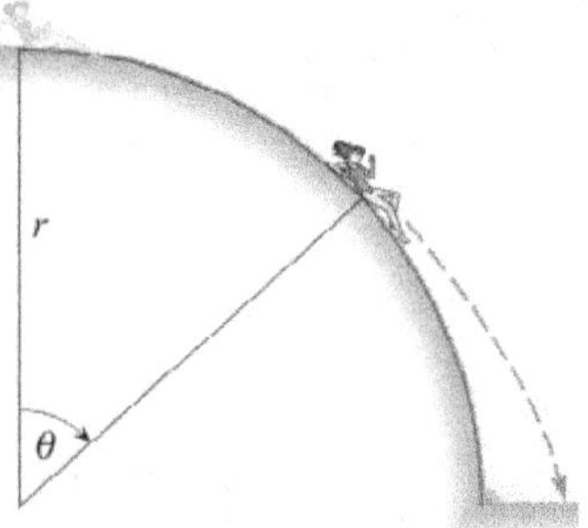

Suppose that water drops are released from a point at the edge of a roof with a constant time interval $\Delta t$ between one water drop and the next. The drops fall a distance h to the ground. If $\Delta t$ is very short ie the number of drops falling though the air at any given instant is very large then the CM of the drops is very nearly at a height (above the ground) of
(A) h/2                                          (B) h/3
(C) 2h/3                                         (D) 3h/4

The moment of inertia of a uniform solid regular tetrahedron of mass m and edge a, about its symmetry axis (i.e. an axis passing through one vertex and the centre of the opposite face) is:
(A) $ma^2/10$                                    (B) $3ma^2/10$
(C) $ma^2/15$                                    (D) $ma^2/20$

A uniform rod lies at rest on a frictionless horizontal surface. A particle, having a mass equal to that of the rod, moves on the surface perpendicular to the rod collides with it, and finally sticks to it.  The minimum loss of KE in the collision, under the given conditions, is
(A) 5%                                           (B) 10%
(C) 20%                                          (D) 30%

8.  A pendulum of length 1m hangs from an inclined wall. Suppose that this pendulum is released at an initial angle of 10° and it bounces off the wall elastically when it reaches an angle of -5° as shown in the figure. Take $g = \pi^2$ m/s². The period of this pendulum is ( in second)
(A) 2/3
(B) 3/2.
(C) 3/4.
(D) 4/3

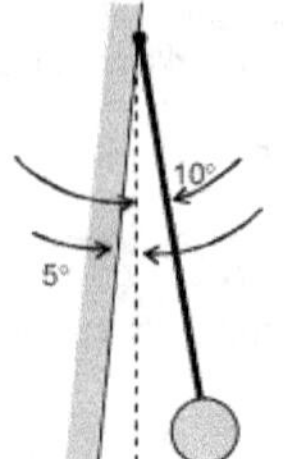


9.  The Earth has a circular orbit of radius r and period t around the Sun; Mars has a circular orbit of radius R and period T. In order to send a spacecraft from the Earth to Mars, it is convenient to launch the spacecraft into an elliptical orbit whose perihelion coincides with the orbit of the Earth and whose aphelion coincides with the orbit of Mars; this orbit requires the least amount of energy for a trip to Mars. The time t' taken by a spacecraft to reach Mars from the Earth satisfies:
(A) $t' = (t + T)/2$
(B) $t'^2 = (t^2 + T^2)/2$
(C) $t'^{3/2} = (t^{3/2} + T^{3/2})/2$
(D) $(2t')^{3/2} = (t^{3/2} + T^{3/2})/2$

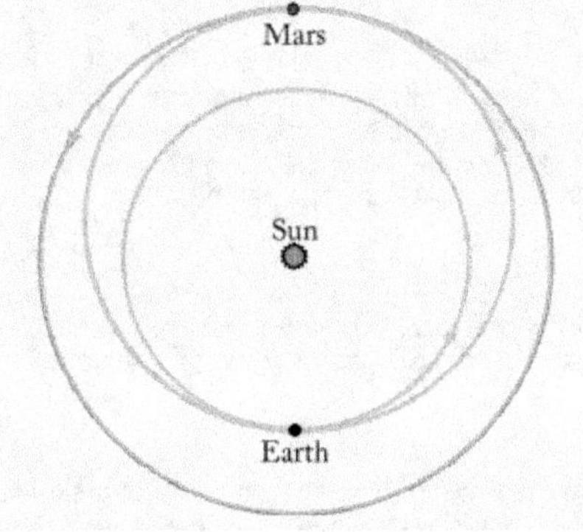


10. When a force accelerates a body immersed in a fluid, some of the fluid must also be accelerated, since it must be pushed out of the way of the body and flow around it. Thus, the force must overcome not only the inertia of the body, but also the inertia of the fluid pushed out of the way. It can be shown that for a spherical body completely immersed in a nonviscous fluid, the extra inertia is that of a mass of fluid half as large as the fluid displaced by the body. The acceleration of a small spherical air bubble in water is nearly
(A) zero.                              (B) g
(C) g/2.                               (D) 2g

A uniform rectangular plate is hanging vertically downward from a hinge that passes along its left edge. By blowing air at 12 m/s over the top of the plate only, it is possible to keep the plate in a horizontal position, as illustrated in part (a) of the drawing. To what value of speed should the air be blown so that the plate is kept at a 30° angle with respect to the vertical, as in part (b) of the drawing?

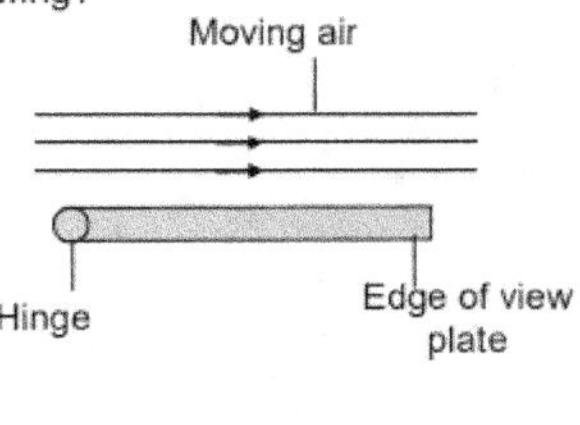


(a)

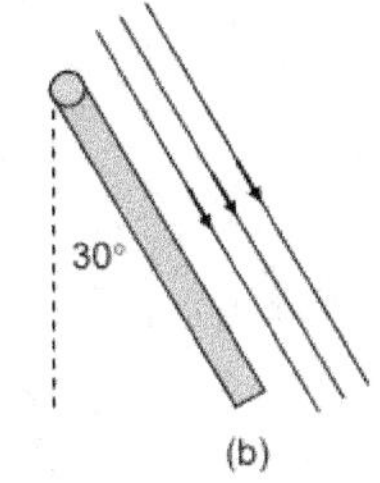


(b)

| | |
|---|---|
| (A) 12m/s | (B) 6 m/s |
| (C) 6√2 m/s | (D) 12√2 m/s |

The drawing shows a bicycle wheel of radius r, resting against a small step whose height is h=r/5. A clockwise torque is applied to the axle of the wheel. As the magnitude of torque increases, there comes a time when the wheel just begins to rise up and loses contact with the ground. Let this torque be τ. What is the magnitude of the horizontal component of the acceleration of the centre of the wheel when a torque of 2τ is applied? Assume that the wheel doesn't slip at the edge of the step when this torque is applied (ignore the mass of the spokes).

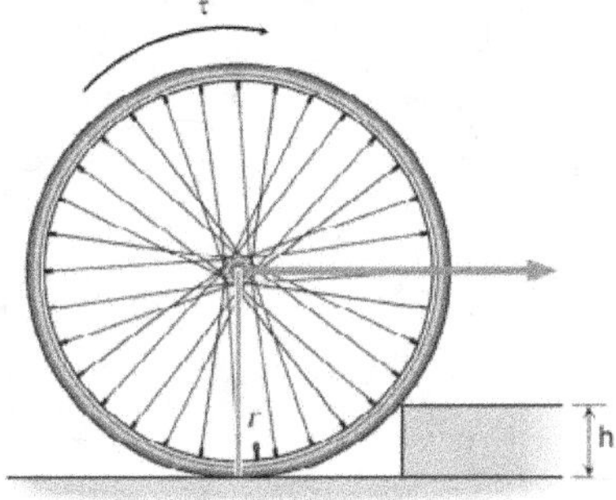


| | |
|---|---|
| (A) 4.8 m/s² | (B) 3 m/s² |
| (C) 2.4 m/s² | (D) 1.5 m/s² |

13. A gas fills the right portion of a horizontal cylinder whose radius is 5.00 cm. The initial pressure of the gas is 101 kPa. A frictionless movable piston separates the gas from the left portion of the cylinder, which is evacuated and contains an ideal spring, as the drawing shows. The piston is initially held in place by a pin. The spring is initially unstrained, and the length of the gas-filled portion is 20.0 cm. When the pin is removed and the gas is allowed to expand, the length of the gas-filled chamber doubles when it finally reaches equilibrium. The initial and final temperatures are equal. Determine the spring constant of the spring.

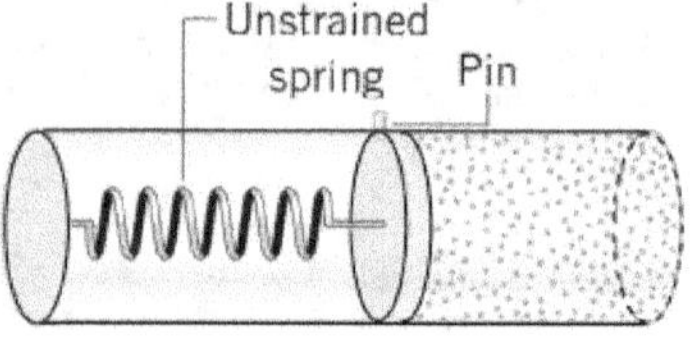


(A) 200 N/m          (B) 20 N/m
(C) 2 N/cm          (D) 2 N/mm

14. Heat flows from a reservoir at 373 K to a reservoir at 273 K through a copper rod as shown in the figure. The heat then leaves the 273 K reservoir and enters a Carnot engine, which uses part of this heat to do work and rejects the remainder to a third reservoir at 173 K. What fraction of the heat leaving the 373 K reservoir is rendered unavailable for doing work, as compared to the situation where a Carnot engine is connected directly between the 373 K and 173 K reservoirs?

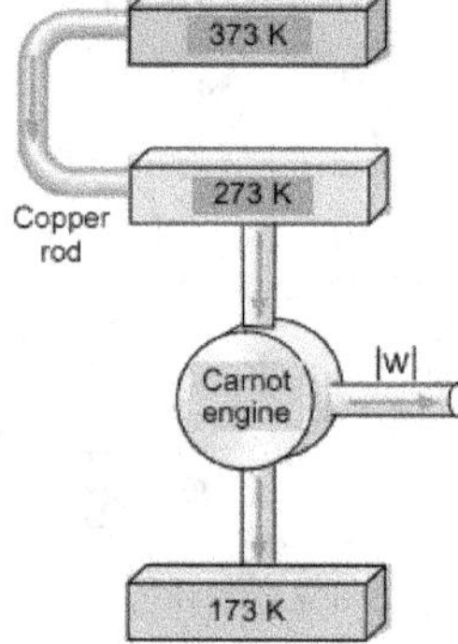


(A) 3%
(B) 10%
(C) 17 %
(D) 35 %

A handclap on stage in an amphitheater sends out sound waves that scatter from terraces of width w = 0.75 m (see figure). The sound returns to the stage as a periodic series of pulses, one from each terrace; the parade of pulses sounds like a played note.  Assuming that all the rays in Figure are horizontal, find the frequency at which the pulses return (that is, the frequency of the perceived note). Take the speed of sound to be 330m/s.

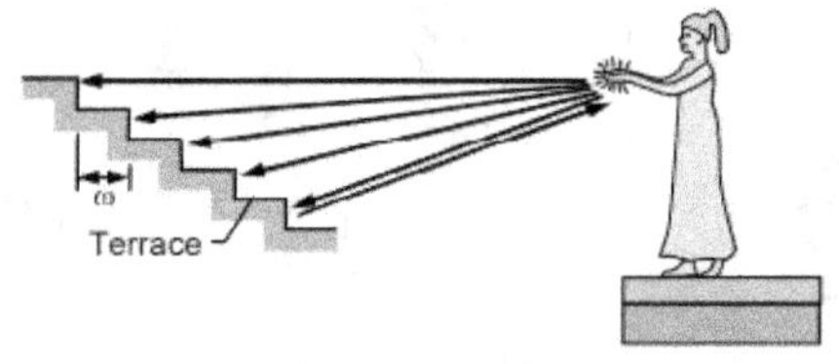


(A) 248Hz
(B) 220 Hz
(C) 456 Hz
(D) 440Hz

An RC- circuit, with R = 600kΩ and C= 10μF, is connected to a 5.0-V battery until the capacitor is fully charged. Then, the battery is suddenly replaced with a new 3.0-V battery of opposite polarity. At what time after this replacement will the energy stored in the capacitor be zero?

(given that $e \approx \dfrac{8}{3}$)

(A) 12 s
(B) 6 s
(C) 3 s
(D)1.5 s

The drawing shows a frictionless incline and pulley.  The two blocks are connected by a wire (mass per unit length, $\mu$ = 25 g/m) and remain stationary.  A transverse wave on the wire has a speed of 60 m/s relative to it. Neglect the weight of the wire relative to the tension in the wire. If the mass $m_2$ be increased by 1%, the speed of the transverse wave will be
(A) 60.2 m/s
(B) 60.3 m/s
(C) 60.4 m/s
(D)60.6 m/s

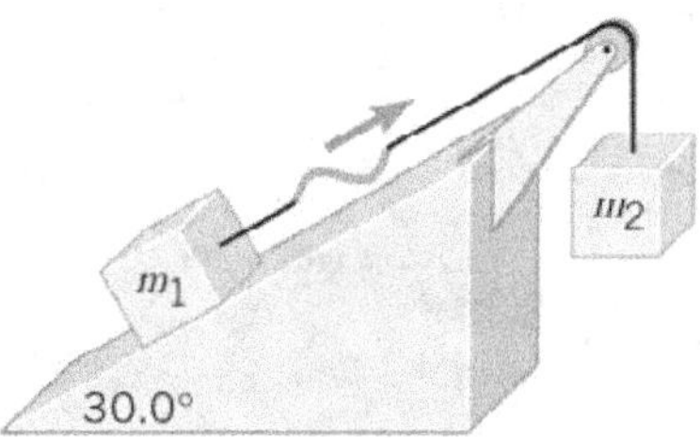

18. The drawing shows a coil of copper wire that consists of two semicircles joined by straight sections of wire. In part (a) the coil is lying flat on a horizontal surface. The dashed line also lies in the plane of the horizontal surface. Starting from the orientation in part (a) the smaller semicircle rotates at an angular frequency $\omega$ about the dashed line, until its plane becomes perpendicular to the horizontal surface, as shown in part (b). A uniform magnetic field B, constant in time and is directed upward, perpendicular to the horizontal surface. The field completely fills the region occupied by the coil in either part of the drawing. The magnitude of the magnetic field is B= 2 T.

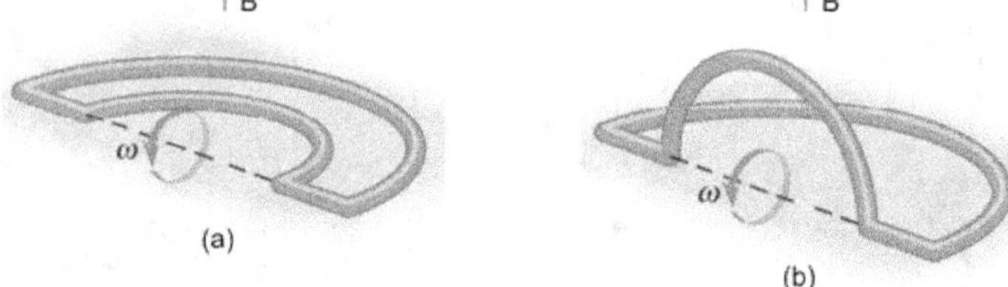


The resistance of the coil is 0.4 $\Omega$, and the smaller semicircle has a radius of 20 cm while the bigger one has a radius of 40 cm. The angular frequency at which the small semicircle rotates is $\omega$ = 15 rad/s. Determine the average current I, if any, induced in the coil as the coil changes shape from that in part (a) of the drawing to that in part ( b).

(A) 0 A.                                    (B) 1A
(C)2A.                                      (D) 3A

19. If a hole is punctured in a tire, the gas inside will gradually leak out of it. Let's assume the following: the area of the hole is A; the tire volume is V; and the time it takes for most of the air to leak out of the tire be t. This time can be expressed in terms of the ratio A/V, the temperature T, the Universal gas constant R, and the mass of the gas molecules inside the tire, m. Under these assumptions, we can use dimensional analysis to find an estimate for t. Assuming this estimate to be correct, if the mass of air within a tyre is increased by 70%, the absolute tyre temperature increased by 20%, while the area of the punctured hole is doubled (but still small) then the time in which a tyre will go flat will

(A)increase by 10%                          (B) increase by 25%
(C) decrease by 25%.                        (D) decrease by 40%

A long wire carries current of 25 A parallel to the positive x axis, except for three of the four segments that follow the edges of a cube of side 0.4 m, as shown in figure. The wire is in a uniform magnetic field of 2.0 T directed parallel to the positive x axis. What is the net magnetic force on the wire (in newton) ?

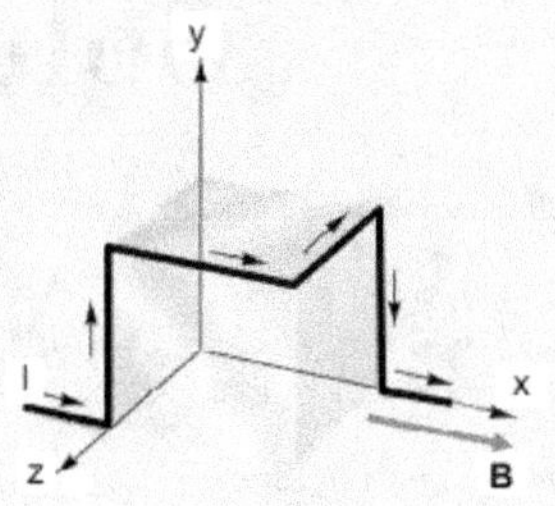


(A) $20\hat{j}$

(B) $20\hat{k}$

(C) $20(\hat{j}-\hat{k})$

(D) $-20\hat{j}$

A uniformly charged square plate having side L carries a uniform surface charge density $\sigma$. The plate lies in the y-z plane with its center at the origin. A point charge q lies on the x-axis. The flux of the electric field of q through the plate is $\phi_0$; while the force on the point charge q due to the plate is $F_0$, along the x-axis. Then,

(A) $\sigma = \dfrac{F_0}{\phi_0 L}$

(B) $\sigma = \dfrac{F_0}{\phi_0}$

(C) $\sigma = \dfrac{F_0 L}{\phi_0}$

(D) $\sigma = \dfrac{\phi_0}{F_0}$

A riverside warehouse has two open doors as shown in Figure. Its walls are lined with sound-absorbing material. A boat on the river sounds its horn. To person A the sound is loud and clear. To person B the sound is barely audible. The principal wavelength of the sound waves is 2 m. Assuming person B is at the position of the first minimum, determine the distance between the doors, center to center. Distance AB = 20 m and the distance of A from the doors is 150 m.

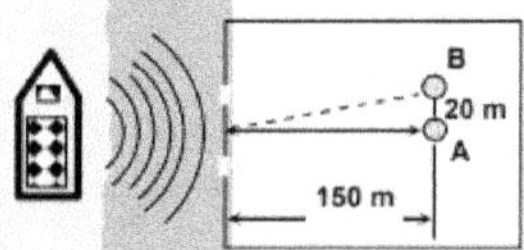


(A) 7.5 m

(B) 15 m

(C) 5 m

(D) 10 m

# Solution

1. $\delta R/R = 2\delta u/u + 2\cot 2\theta\ \delta\theta$

2. If s be their separation at time t then,
$$s^2 = (0.4-80t)^2 + (0.6-60t)^2$$
where t is in hour, s in km. Minimise s using calculus.

3. The horizontal acceleration of Block 1 is always greater than that of block 2.

4. The normal reaction vanishes when the person loses contact with the surface:
$$mv^2/r = mg\cos\theta;$$
Conservation of energy gives:
$$1/2\ mv^2 + mgr\cos\theta = 1/2\ mu^2 + mgr; \text{ where } u^2 = 0.5gr$$

5. The droplets fall a distance $\frac{1}{2}gt^2$ in time t, and the number of the droplets and hence their mass is proportional to dt. Computing the CM of the droplets using the definition, we get the result.

6. For an equilateral triangle of side a, moment of inertia is $\dfrac{ma^2}{12}$ about an axis passing through its CM and perpendicular to plane - the result of the integration is similar to that of a solid cone along its axis: a factor of 3/5.
$$I = \frac{3}{5}\left(\frac{ma^2}{12}\right)$$

7. The maximum loss occurs when the collision is "head-on" and the minimum is when it collides at an end. Assume that the particle strikes the rod (of length 2L) perpendicularly at a distance x from its centre. Apply conservation of momentum and angular momentum to calculate the final velocity of the rod and also its angular velocity. The fractional loss in KE $= 1/[3(x/L)^2 + 2]$

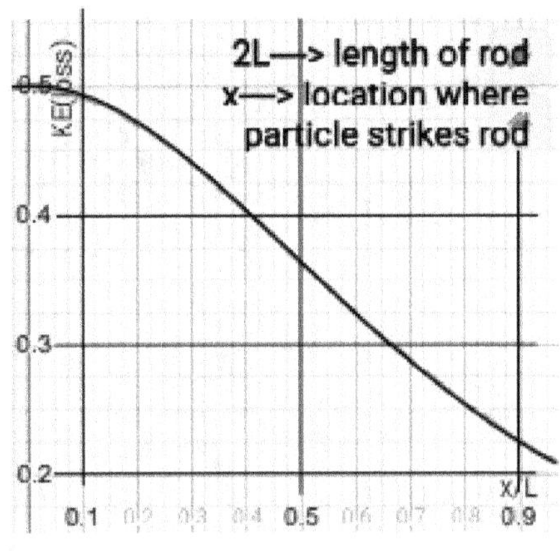


8. We take $\theta = (10°)\cos(\pi t)$; calculate t when $\theta = -5°$. This gives t = 2/3 (s). The period is 2 times this.

9. Use Kepler's Law of Periods.

If m be the mass of water displaced, the buoyant force is mg, while the mass of the extra water that has to be pushed "out of the way" is effectively m/2.

Taking torques about the hinge, we get:
In the 1st case, $mgL/2 = p\ AL/2$
In the second case, $mgL/4 = p'\ AL/2$

When the wheel gets just lifted $\tau = mg(3r/5)$. But when a torque of $2\tau$ is applied, we can write:
$mg(3r/5) = I\alpha$,
where $I = 2mr^2$.
The horizontal acceleration is $r\alpha(4/5)$.

The final pressure is $p = \dfrac{101}{2}\,kPa$ ; $kx = p\pi r^2$ where $x = 0.2$, the compression in the spring and $r = 0.05$ the radius of cylinder.

With reservoirs at 273 K, 173 K:
$$\frac{W}{Q} = \frac{273 - 173}{273} = 0.37$$
With reservoirs at 373 K, 173 K:
$$\frac{W'}{Q} = \frac{373 - 173}{373} = 0.54$$
$W' - W = (0.54 - 0.37)Q = 0.17\ Q$

$$T = \frac{2w}{v}\ ;\ f = \frac{1}{T}$$

$$\frac{3\ volt}{(3 + 5)volt} = e^{-t/\tau}\ ;\ \tau = RC = 6\ s$$
$\therefore t \approx 6$ s.

Tension, $T = \dfrac{3m_1 m_2}{2(m_1 + m_2)}g$ ; $v = \sqrt{\dfrac{T}{\mu}}$ ; $\dfrac{\delta v}{v} = \dfrac{1}{2}\dfrac{\delta T}{T} = \dfrac{1}{3}\dfrac{\delta m_2}{m_2}$ (where, $m_1 \approx 2m_2$)

The change in flux $\Delta\phi = \dfrac{\pi}{2}r^2 B$ , where $r = 0.2$ m

Average emf = $\dfrac{\Delta\phi}{\pi/2\omega}$

Current = average emf / resistance

Using dimensional analysis or otherwise $t = \dfrac{V}{A}\sqrt{\dfrac{m}{RT}} \times$ some numerical factor

Substituting the values we get $\dfrac{t'}{t} = \dfrac{1}{2}\sqrt{\dfrac{1.7}{1.2}} \approx 0.6$

Force $= 25(0.4\hat{i} - 0.4\hat{k}) \times 2\hat{i} = -20\hat{j}$

The correct answer can be determined by dimensional analysis.

22. $\dfrac{\lambda D}{2d} = AB = 20$ m

where $\lambda = 2$ m
$D = 150$ m
$\therefore d = 7.5$ m

23. From geometry $\beta = (180° - 2\theta)$

So, $\dfrac{d\beta}{dt} = -\dfrac{2d\theta}{dt}$

24. When the switch closes at $V_c = 2V/3$ the capacitor discharges through $R_2 = 3R$

T(discharge) = $3RC \, \ell n \, 2$ (since the voltage halves).

The charging occurs through $R_1 + R_2 = 9R$

T (charge) = $9RC \ell n2$

$T = 12 \, RC \, \ell n \, 2 \approx 8.4 \, RC$

25. When the left end is positive, upper diode conducts and lower diode is cut off.

$R_{eq} = R$

When right end is positive the lower diode conducts.

$R_{eq} = R + \dfrac{3R}{4} = \dfrac{7R}{4}$

$\therefore$ Power $= \dfrac{(\Delta V)^2}{R}\left[1 + \dfrac{4}{7}\right] \times \dfrac{1}{2} \approx \dfrac{(\Delta V)^2}{R} \times 0.8$

26. The object must be located at the centre of the curvature of the mirror for this to happen.
For the inverted image formed directly by the lens we can write

$\dfrac{1}{v} - \dfrac{1}{u} = \dfrac{1}{10} , \ \dfrac{v}{u} = -\dfrac{3}{2}$

27. This occurs when the angle of incidence is equal to the Brewster angle: $\tan \theta_B = \mu$.

28. If the sphere has a uniform mass density (total mass m), then $\dfrac{\mu}{L} = \dfrac{q}{2m}$, where $L = \dfrac{2}{5}mR^2\omega$.

29. $h\nu = E_{n+1} - E_n = E_0\left(\dfrac{1}{n^2} - \dfrac{1}{(n+1)^2}\right) \approx \dfrac{E_0 2n}{n^4}$, for large n.

30. $m_n = \dfrac{m_1 v_1 - m_2 v_2}{v_2 - v_1}$

$= \dfrac{14 \times (4.7 \times 10^6) - 1 \times (3.3 \times 10^7)}{3.3 \times 10^7 - 4.7 \times 10^6} = 1.16$

# Mock Test 5

A transmitter supplies 9 kw to the aerial when unmodulated. The power radiated when modulation index is 60%, is

(A) 9.72 kw  (B) 9 kw
(C) 5.4 kw  (D) 10.62 kw

The combination of the gates shown in figure produces

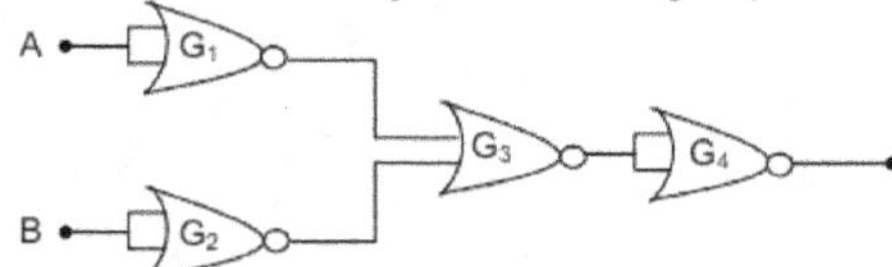

(A) AND gate  (B) XOR gate
(C) NOR gate  (D) NAND gate

Two radioactive materials $X_1$ and $X_2$ have decay constants $10\lambda$ and $\lambda$ respectively. If initially they have the same number of nuclei, then the ratio of number of nuclei of $X_1$ to that of $X_2$ will be $\dfrac{1}{e}$ after time

(A) $\dfrac{1}{10\lambda}$  (B) $\dfrac{1}{11\lambda}$

(C) $\dfrac{1}{9\lambda}$  (D) $\dfrac{11}{10\lambda}$

4.      When the hydrogen atom is raised from the ground state to fifth state
        (A) both K.E and P.E increases                (B) both K.E and P.E decreases
        (C) P.E increases and K.E decreases           (D) P.E decreases and K.E increases

5.      In an ideal parallel LC circuit, the capacitors is charged by connecting it to a dc source which is
        then disconnected. The current in the circuit
        (A) becomes zero instantaneously              (B) grows monotonically
        (C) decays monotonically                      (D) oscillates instantaneously.

6.      Which of the following graph is applicable between $\dfrac{1}{v}$ and $\dfrac{1}{u}$ for a concave mirror

(A) 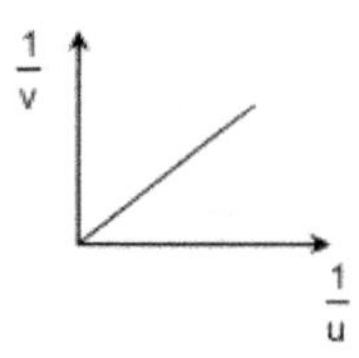

(B) 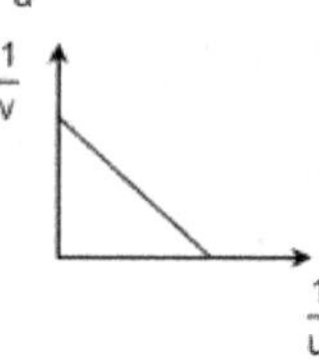

(C) 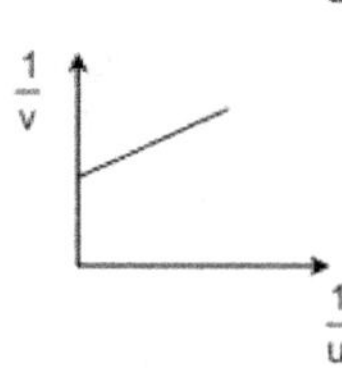

(D) 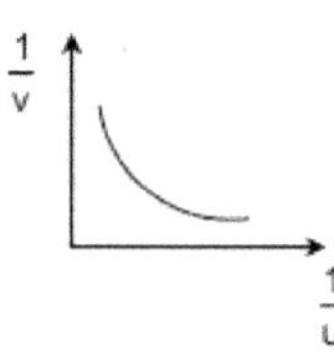

7.      In a voltage regulating of zener diode, the graph of output voltage $v_o$ versus input voltage $v_i$ is

(A) 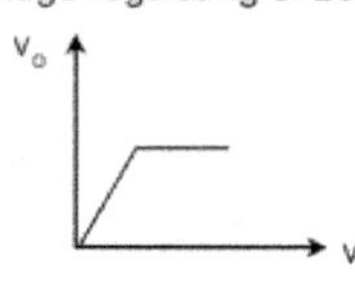

(B) 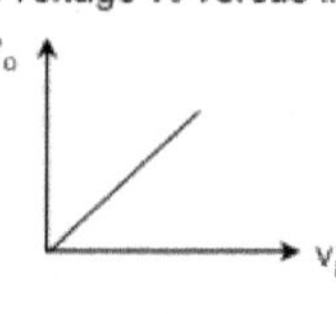

(C) 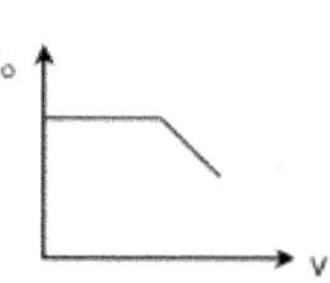

(D) 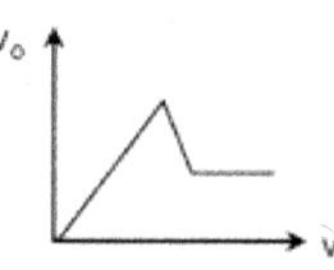

From ideal instruments, current measured is $i = 10.0$ amp, potential difference measured is $V = 100.0$ volt, length of wire is 31.4 cm, and diameter of wire is 2.00 mm, the resistivity of wire will be (in correct significant figures) ($\pi = 3.14$)

(A) $1.00 \times 10^{-4} \Omega - m$

(B) $1.0 \times 10^{-4} \Omega - m$

(C) $1 \times 10^{-4} \Omega - m$

(D) $1.000 \times 10^{-4} \Omega - m$

A cube of mass m and side a is moving along a plane with constant speed $v_0$ as shown in figure. The magnitude of angular momentum of the cube about z -axis would be.

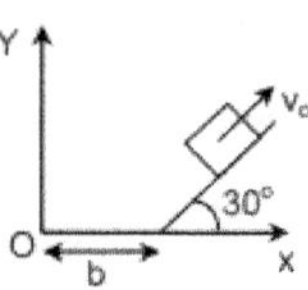

(A) $\dfrac{mv_0 b}{2}$

(B) $\dfrac{\sqrt{3}mv_0 b}{2}$

(C) $mv_0 \left( b - \dfrac{a}{2} \right)$

(D) None of these

A disc of radius R is rolling purely on a flat horizontal surface, with constant angular velocity. The angle between the velocity and acceleration vectors of point P is

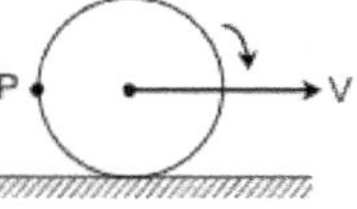

(A) zero

(B) $45°$

(C) $135°$

(D) $\tan^{-1}\left(\dfrac{1}{2}\right)$

A small mass slides down an inclined plane of inclination $\theta$ with the horizontal. The coefficient of friction is $\mu = \mu_0 x$ where x is the distance through which the mass slides down and $\mu_0$ is constant. Then the distance covered by the mass before it stops is

(A) $\dfrac{2}{\mu_0} \tan \theta$

(B) $\dfrac{4}{\mu_0} \tan \theta$

(C) $\dfrac{1}{2\mu_0} \tan \theta$

(D) $\dfrac{1}{\mu_0} \tan \theta$

16. A heavy but uniform rope of length L is suspended from a ceiling. A particle is dropped from the ceiling at the instant when the bottom end is given a transverse wave pulse. Where will the particle meet the pulse

(A) at a distance $\dfrac{2L}{3}$ from the bottom

(B) at a distance $\dfrac{L}{3}$ from the bottom

(C) at a distance $\dfrac{3L}{4}$ from the bottom

(D) None of the above

17. A particle moves according to the law $x = a \cos \dfrac{\pi t}{2}$. The distance covered by it in the time internal between t = 0 to t = 3 sec is.

(A) 2a

(B) 3a

(C) 4a

(D) a

18. Two concentric spherical shells are as shown in figure. The V – r graph will be as

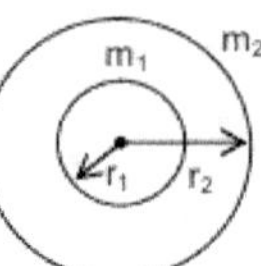

(A) 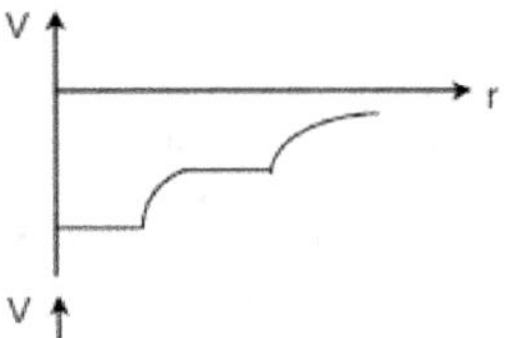

(B) 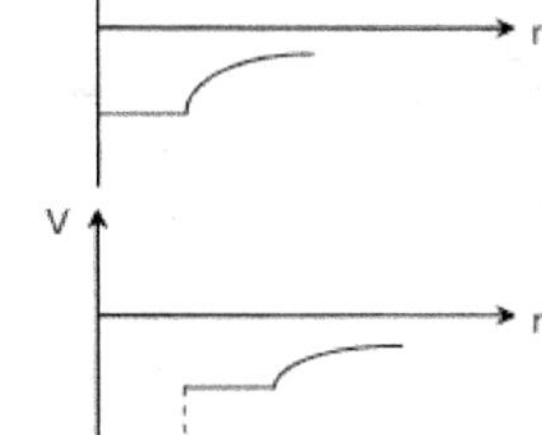

(C)

(D)

Acceleration of a particle moving in x-y plane varies with time t as. $\vec{a} = \left( t\,\hat{i} + 3t^2\,\hat{j} \right)$.

Here a is in m/s² and t in sec. At time t = 0 particle is at rest at origin. Mass of the particles is 1 kg. Find the net work done on the particle in first 2 sec.

(A) 40 J          (B) 34 J

(C) 16 J          (D) 48 J

Both the blocks as shown in figure are given together a horizontal velocity towards right. The acceleration of the centre of mass of the system of blocks is $(m_A = 2m_B = 2\text{kg})$.

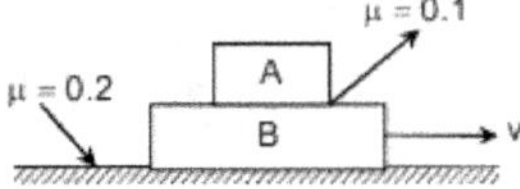


(A) zero

(B) $\dfrac{5}{3}\,m/s^2$

(C) $\dfrac{7}{3}\,m/s^2$

(D) $2\ m/s^2$

Figure here shows the vertical cross section of a vessel filled with a liquid of density ρ. The normal thrust per unit area due to liquid on the walls of the vessel at point A as shown, will be

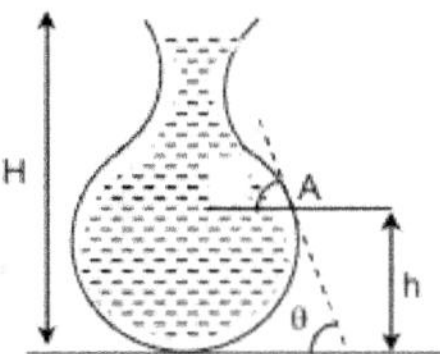


(A) hρg

(B) Hρg

(C) $(H-h)\rho g$

(D) $(H-h)\rho g \cos\theta$

The volume thermal expansion coefficient of an ideal gas at constant pressure is

(A) T          (B) $T^2$

(C) $\dfrac{1}{T}$          (D) $\dfrac{1}{T^2}$

19. Two coherent point sources $S_1$ and $S_2$ vibrating in phase emit light of wavelength $\lambda$. The separation between the sources is $2\lambda$. The smallest distance from $S_2$ on a line passing through $S_2$ and perpendicular to $S_1S_2$, where a minimum of intensity occurs is

(A) $\dfrac{7\lambda}{12}$

(B) $\dfrac{15\lambda}{4}$

(C) $\dfrac{\lambda}{2}$

(D) $\dfrac{3\lambda}{4}$

20. Light is incident normally on face AB of a prism as shown in figure. A liquid of refractive index $\mu$ is placed on face AC of the prism. The prism is made of glass of refractive index $\dfrac{3}{2}$. The limits of $\mu$ for which total internal reflection takes place on face AC is

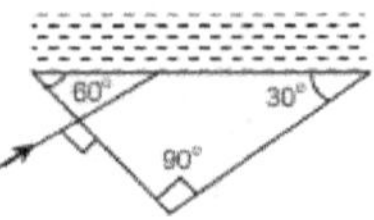


(A) $\mu > \dfrac{\sqrt{3}}{2}$

(B) $\mu < \dfrac{3\sqrt{3}}{4}$

(C) $\mu > \sqrt{3}$

(D) $\mu < \dfrac{\sqrt{3}}{2}$

21. The image of point P when viewed from top of the slabs will be
(A) 2 cm above P
(B) 1.5 cm above P
(C) 2 cm below P
(D) 1 cm below P

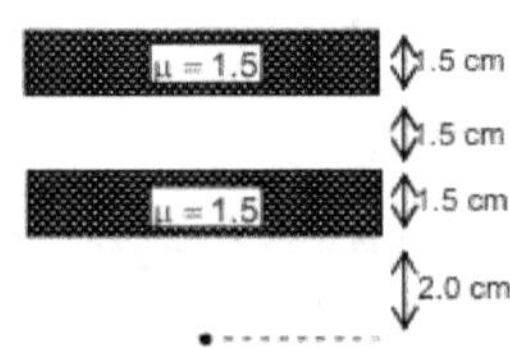

. If the capacitance of each capacitor is C, then effective capacitance of the shown network across junction ab is

(A) 2 C

(B) C

(C) $\dfrac{C}{2}$

(D) 5 C

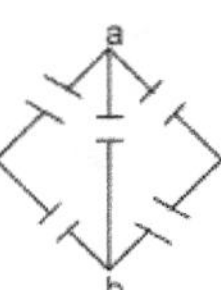

. For given circuit charge on capacitor C, and $C_2$ in steady state will be equal to

(A) $C_1(V_A - V_C), C_2(V_C - V_B)$ respectively

(B) $C_1(V_A - V_B), C_2(V_A - V_B)$ respectively

(C) $(C_1 + C_2)(V_A - V_B)$ on each capacitor

(D) $\left(\dfrac{C_1 C_2}{C_1 + C_2}\right)(V_A - V_B)$ on each capacitor

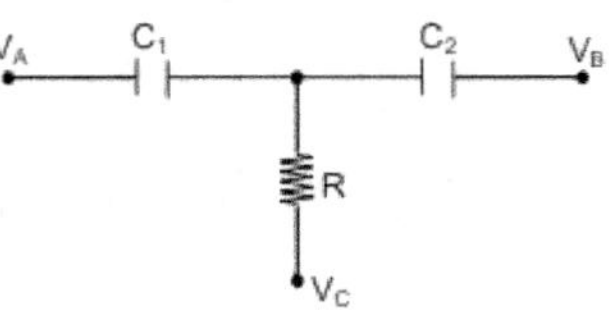

. In the diagram resistance between any two junctions is R. Equivalent resistance across terminals A and B is (Assume P and Q across as a junction)

(A) $\dfrac{11\,R}{7}$

(B) $\dfrac{18\,R}{11}$

(C) $\dfrac{7\,R}{11}$

(D) $\dfrac{11R}{18}$

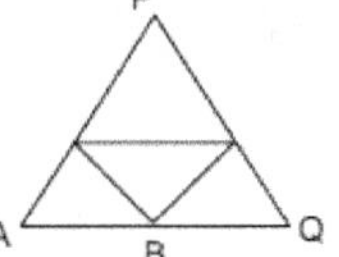

. A particle of charge per unit mass $\alpha$ is released from origin with velocity $\vec{V} = v_o\hat{i}$ in a magnetic field. $\vec{B} = -B_o\hat{K}$ for $x \le \dfrac{\sqrt{3}}{2}\dfrac{v_o}{B_o\alpha}$ and $\vec{B} = 0$ for $x > \dfrac{\sqrt{3}}{2}\dfrac{v_o}{B_o\alpha}$

and $\vec{B} = 0$ for $x > \dfrac{\sqrt{3}}{2}\dfrac{v_o}{B_o\alpha}$

The x – coordinate of the particle at time $t > \left(\dfrac{\pi}{3B_o\alpha}\right)$ would be

(A) $\dfrac{\sqrt{3}}{2}\dfrac{v_o}{B_o\alpha} + \dfrac{\sqrt{3}}{2}v_o\left(t - \dfrac{\pi}{B_o\alpha}\right)$

(B) $\dfrac{\sqrt{3}}{2}\dfrac{v_o}{B_o\alpha} + v_o\left(t - \dfrac{\pi}{3B_o\alpha}\right)$

(C) $\dfrac{\sqrt{3}}{2}\dfrac{v_o}{B_o\alpha} + \dfrac{v_o}{2}\left(t - \dfrac{\pi}{3B_o\alpha}\right)$

(D) $\dfrac{\sqrt{3}}{2}\dfrac{v_o}{B_o\alpha} + \dfrac{v_o t}{2}$

# Solution

1. **D**

$$P_t = P_c\left(1 + \frac{m^2}{2}\right) = 9\left(1 + \frac{(0.6)^2}{2}\right) = 9\left[1 + \frac{0.36}{2}\right]$$

$$= 9[1 + 0.18] = 9[1.18] = 10.62 \text{ kw}$$

2. **D**

Out put of $G_1 = \bar{A}$, out put of $G_2 = \bar{B}$

Out put of $G_3 = AB$

Now $Y = \overline{AB}$ (NAND gate)

3. **C**

$$N_1 = N_0 e^{-10\lambda t}, N_2 = N_0 e^{-\lambda t} \therefore \frac{N_1}{N_2} = \frac{1}{e^{9\lambda t}}$$

$$\therefore \frac{1}{e} = \frac{1}{e^{9\lambda t}} \therefore t = \frac{1}{9\lambda}$$

4. **C**

As the number of orbit increases the velocity decreases i.e K.E decreases the potential energy becomes less negative i.e P.E increase.

5. **C**

The current in the circuit decays monotonically.

6. **B**

$$\frac{1}{v} + \frac{1}{u} = \frac{1}{f}$$

Therefore, for a given mirror $\frac{1}{v}$ versus $\frac{1}{u}$ graph should be a straight line at u = f or

$$\frac{1}{u} = \frac{1}{f}, v = \infty \text{ or } \frac{1}{v} = 0 \text{ and viceversa}$$

7. **A**

Fact

8. **C**

$$\rho = \frac{\pi D^2}{4L}\frac{v}{i} = \frac{(3.14)\left(2.00 \times 10^{-3}\right)^{-2}}{4(0.314)}\left(\frac{100.0}{10.0}\right)$$

$$\rho = 1.00 \times 10^{-4} \Omega - m$$

**D**

$$r_2 = (b-a)\sin 30^\circ$$

$$= \left(\frac{b-a}{2}\right)$$

$$\therefore L = mv_o r_2 = \frac{mv_o(b-a)}{2}$$

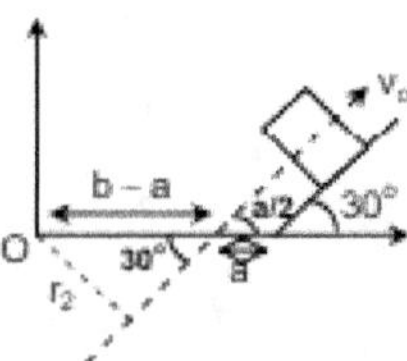


**B**

Angle is $45^\circ$

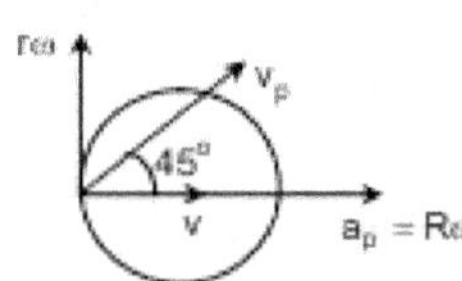


**A**

$$a = g\sin\theta - (\mu_o g\cos\theta)x \quad \therefore \int_0^v v\,dv = \int_0^{x_{max}}\left[g\sin\theta - (\mu_o g\cos\theta)x\right]dx$$

After solving we get $x_{max} = \dfrac{2\tan\theta}{\mu_o}$

**B**

$$\frac{d\vec{v}}{dt} = t i + 3t^2 j \quad \therefore \int_0^v d\vec{v} = \int_0^2 \left(t i + 3t^2 j\right)dt$$

$$\therefore \vec{v} = (2i + 8j)m/s.$$

$$\therefore v = \sqrt{68}$$

by work energy theorem $\therefore w = \dfrac{1}{2}mv^2 \therefore w = 34J$

**D**

Net external force on the two blocks (whether they move with same retardation or not)

$$f_{ext} = (0.2)(2+1)\times 10 = 6N$$

$$\therefore a_{cm} = \frac{f_{ext}}{2+1} = 2\ m/s^2$$

**C**

Normal thrust per unit area is actually the pressure due to liquid

**C**

$$PV = nRT \therefore pdv = nRdT \therefore dv = \left(\frac{v}{T}\right)dT \qquad .....(i)$$

$$dv = \gamma v dT \qquad .......(ii)$$

from (i) & (ii) $\gamma = \dfrac{1}{T}$

16. **B**

$$v = \sqrt{gh} \therefore \frac{dh}{dt} = \sqrt{gh} \therefore t = \int_0^h \frac{dh}{\sqrt{gh}} \text{ or } t = 2\sqrt{\frac{h}{g}}$$

Now at the time of meeting,
Time of fall of particle = time of wave pulse on reaching up to these

$$\sqrt{\frac{2(L-h)}{g}} = 2\sqrt{\frac{h}{g}} \therefore h = \frac{L}{3}$$

17. **B**

Comparing the given equation with $x = a\cos\omega t$

$$\therefore \omega = \frac{\pi}{2} \text{ or } \frac{2\pi}{T} = \frac{\pi}{2} \therefore T = 4\,sec$$

The given time $t = 3$ sec is really $\dfrac{3T}{4}$.

$\therefore$ distance covered will be 3 a

18. **C**

for $r \le r_1$ $V = \dfrac{Gm_1}{r_1} - \dfrac{Gm_2}{r_2} = constant$

for $r_1 \le r \le r_2$ $V = -\dfrac{Gm_2}{r_2} - \dfrac{Gm_1}{r_1}$

slope of V - x graph $dV/dr = Gm_1 / r^2$

for $r \ge r_2$ $V = -\dfrac{G}{r}(m_1 + m_2)$

slope of V - r graph $= \dfrac{G}{r}(m_1 + m_2)$

At the boundary of outer shell slope of V – r graph changes from

$\dfrac{Gm_1}{r_2^2}$ to $\dfrac{G(m_1 + m_2)}{r_2^2}$ i.e slope increses

19. **A**

Path difference at $S_2$ is $2\lambda$. Therefore for minimum intensity at P.

$$S_1P - S_2P = \frac{3\lambda}{2} \qquad \ldots\ldots(i)$$

or $\sqrt{4\lambda^2 + x^2} - x = \dfrac{3\lambda}{2}$

Solving equation, $x = \dfrac{7\lambda}{12}$

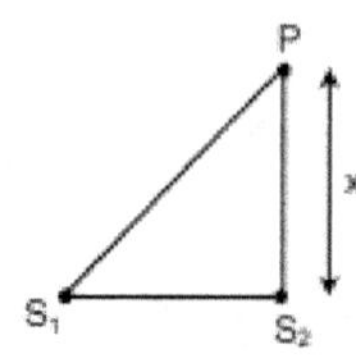


20. **B**

Critical angle between glass and liquid face is $\sin\theta_C = \dfrac{\mu}{3/2}$

$\therefore \sin\theta_C = \dfrac{2\mu}{3}$

Angle of incidence at face AC is $60°$
For TIR to take plane $I > \theta_C$

Or $\sin 60° > \dfrac{2\mu}{3}$ or $\mu < \dfrac{3\sqrt{3}}{4}$.

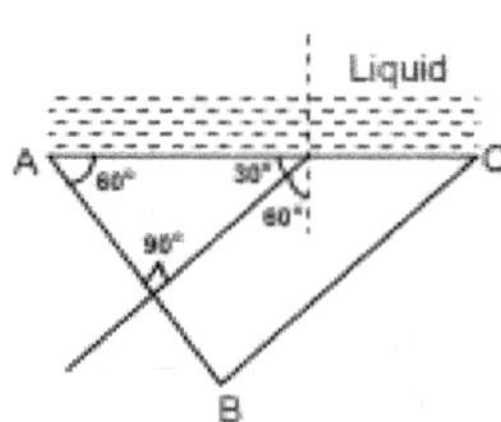

1. **D**

The two slabs will shift the image a distance

$$d = 2\left(1 - \frac{1}{\mu}\right)t = 2\left(1 - \frac{1}{1.5}\right)(1.5) = 1\ cm$$

2. **B**

The given network is a balanced Wheatstone bridge with one capacitor in parallel with this bridge.

3. **A**

No current will flow through the resistance at steady state.

4. **D**

Series and parallel grouping of resistors.

5. **C**

$$r = \frac{mv_o}{B_o q} = \frac{v_o}{B_o \alpha}$$

$$\frac{x}{r} = \frac{\sqrt{3}}{2} = \sin\theta \therefore \theta = 60^\circ$$

$$t_{OA} = \frac{T}{6} = \frac{\pi}{3B_o \alpha}$$

Therefore $x$ – coordinate of particle at any time $t > \dfrac{\pi}{3B_o \alpha}$ will be

$$x = \frac{\sqrt{3}}{2}\frac{v_o}{B_o \alpha} + v_o\left(t - \frac{\pi}{3B_o \alpha}\right)\cos 60^\circ \quad So \quad x = \frac{\sqrt{3}}{2}\frac{v_o}{B_o \alpha} + \frac{v_o}{2}\left(t - \frac{\pi}{3B_o \alpha}\right)$$

6. **A**

$$B = 2\left[\frac{\mu_o}{4\pi}\cdot\frac{i}{d/\sqrt{2}}\right](\sin 90^\circ - \sin 45^\circ)$$

$$B = \frac{\mu_o i}{\sqrt{2}\pi d}\left(1 - \frac{1}{\sqrt{2}}\right)\otimes$$

7. **B**

Polarity of emf will be opposite in the two cases while entering and while leaving the coil only in option (b) polarity is changing

8. **C**

The equivalent circuit is as shown in fig.

$$U_{L(max)} = U_{C(max)} = \frac{1}{2}\frac{(3Q)^2}{3C} = \frac{3Q^2}{2C}$$

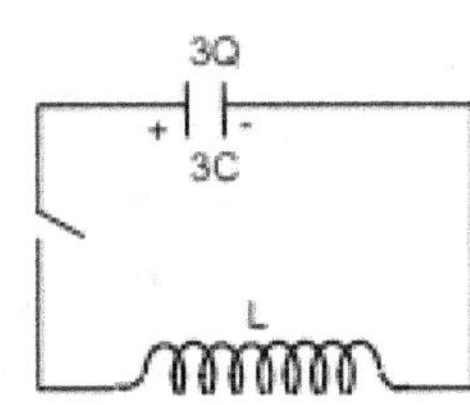

29. **D**

A solid with completely filled valence band is an insulator, if the energy gap between the valence band and the empty conduction band is larger than about 5 eV

30. **D**

$$A = 2\pi Rh = 2(\pi)\left(6.4 \times 10^6\right)(150)m^2$$

$$= 1.96\pi \times 10^3 km^2$$

# Mock Test 6

A particle moves along a circle of radius 'R' with speed varies as $v = a_0 t$, where $a_0$ is a positive constant. Then the angle between the velocity vector and the acceleration vector of the particle when it has covered one fourth of the circle is

(A) $\tan^{-1}\left(\dfrac{\pi}{4}\right)$

(B) $\tan^{-1}\left(\dfrac{\pi}{2}\right)$

(C) $\tan^{-1}(\pi)$

(D) $\tan^{-1}(2\pi)$

Two blocks A and B of masses 2 kg and 8 kg respectively are kept in contact and pressed against a rough vertical wall by applying a horizontal force F = 50 N as shown. The frictional force acting between the two blocks is
(A) zero
(B) 6N
(C) 8 N
(D) 10 N

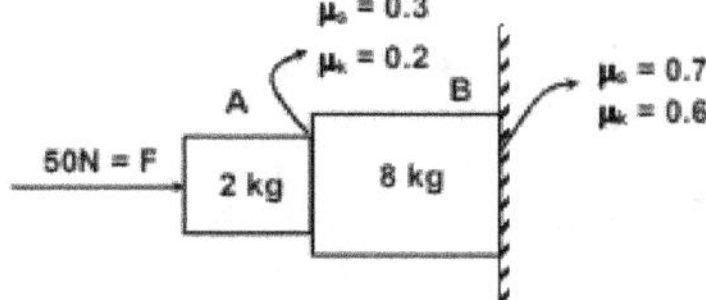


A small body of mass 'm' is placed at the top of a smooth sphere of radius 'R' placed on a horizontal surface as shown. Now the sphere is given a constant horizontal acceleration $a_0 = g$. The velocity of the body relative to the sphere at the moment when it loses contact with the sphere is

(A) $\sqrt{\dfrac{gR}{3}}$

(B) $\sqrt{\dfrac{2gR}{3}}$

(C) $\sqrt{\dfrac{5gR}{3}}$

(D) $\sqrt{\dfrac{7gR}{3}}$

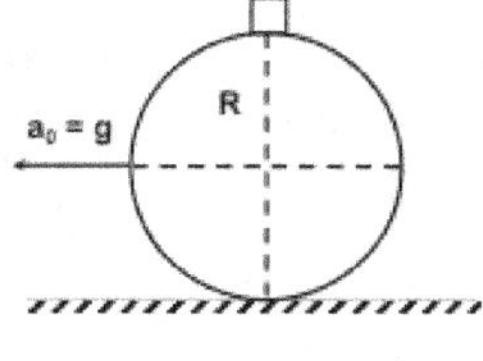


A ball of mass m is attached to the lower end of a vertical string whose upper end is fixed. An identical ball 'B' moving with a velocity $v_0 = 6$ m/s at an angle $\theta = 45°$ from vertical collides elastically with the ball 'A' as shown. The velocity of the ball 'B' just after collision is
(A) 5 m/s
(B) 4 m/s
(C) 3 m/s
(D) 2 m/s

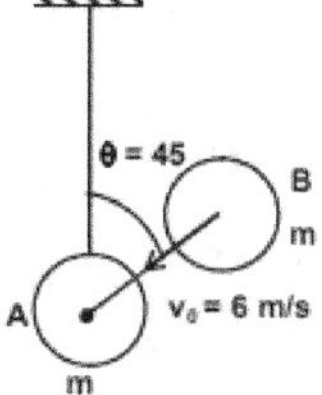


A thin uniform spherical shell of mass m = 2 kg and radius R = 10 cm is placed on a rough horizontal surface with a coefficient of friction $\mu$ = 0.4. A horizontal force 'F' is applied to the spherical shell at a height R/3 above the centre as shown. The maximum magnitude of the applied force 'F' for which the spherical shell will be rolling without slipping is (take $g = 10$ m/s²)
(A) 20 N
(B) 30 N
(C) 40 N
(D) 60 N

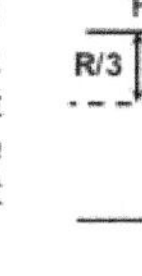


A block of mass 2 kg is kept on a truck moving with a constant acceleration. The height 'h' of the block is 60 cm and its width 'w' is 15 cm. The surface between the block and the truck is sufficiently rough to prevent slipping. The minimum acceleration of the truck at which the block will topple is (take $g = 10$ m/s²)
(A) 2.5 m/s²
(B) 3.5 m/s²
(C) 4.5 m/s²
(D) 5.5 m/s²

7. A uniform solid sphere of mass m and radius 'R' is imparted an initial velocity $v_0$ and angular velocity $\omega_0 = \dfrac{2v_0}{R}$ and then placed on a rough inclined plane of inclination '$\theta$' and coefficient of friction $\mu = 2\tan\theta$ as shown. The time after which the sphere will start rolling without slipping is

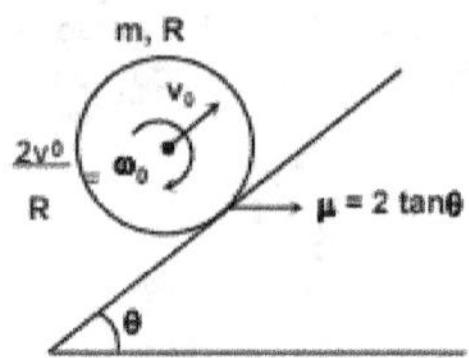



(A) $\dfrac{v_0}{2g\sin\theta}$

(B) $\dfrac{v_0}{4g\sin\theta}$

(C) $\dfrac{v_0}{6g\sin\theta}$

(D) $\dfrac{v_0}{8g\sin\theta}$

8. Two thin films of same liquid of surface tension 'T' are formed between a smooth rectangular wire frame and two thin uniform straight wires each of mass 'm' and length 'l' connected to a massless non-deformed spring of stiffness 'k' and then the system is released from rest. The maximum elongation produced in the spring is

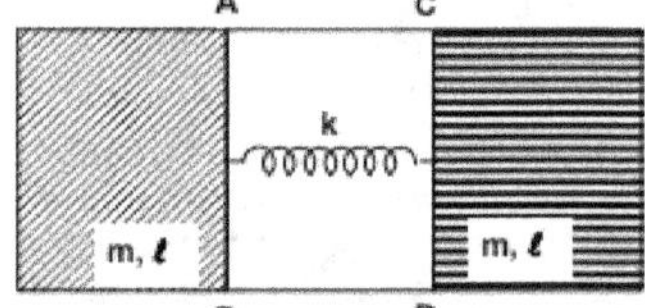



(A) $\dfrac{2T\ell}{K}$

(B) $\dfrac{4T\ell}{K}$

(C) $\dfrac{6T\ell}{K}$

(D) $\dfrac{8T\ell}{K}$

9. A steel wire of negligible mass, length 2l, cross sectional area 'A' and Young's modulus 'Y' is held between two rigid walls. A small body of mass 'm' is suspended from its middle point 'O' as shown. The vertical descent of the middle point 'O' of the wire is

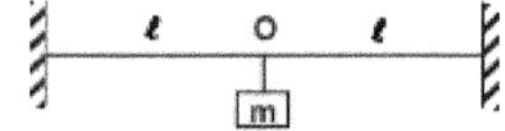



(A) $\ell\left(\dfrac{mg}{AY}\right)^{1/3}$

(B) $\ell\left(\dfrac{2mg}{AY}\right)^{1/3}$

(C) $\ell\left(\dfrac{4mg}{AY}\right)^{1/3}$

(D) $\ell\left(\dfrac{6mg}{AY}\right)^{1/3}$

A tunnel is dug across the earth of mass 'M' and radius 'R' at a distance R/2 from the centre 'O' of the Earth as shown. A body is released from rest from one end of the smooth tunnel. The velocity acquired by the body when it reaches the centre 'C' of the tunnel is

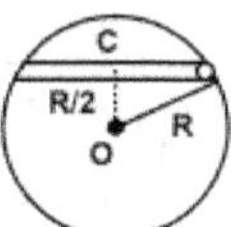

(A) $\sqrt{\dfrac{3GM}{2R}}$

(B) $\sqrt{\dfrac{3GM}{4R}}$

(C) $\sqrt{\dfrac{3GM}{8R}}$

(D) $\sqrt{\dfrac{3GM}{16R}}$

A uniform circular disc of mass 'm' and radius 'R' is placed on a rough horizontal surface and connected to the two ideal non-deformed springs of stiffness k and 2k at the centre 'C' and point 'A' as shown. The centre of the disc is slightly displaced horizontally from equilibrium position and then released, then the time period of small oscillation of the disc is (there is no slipping between the disc and the surface).

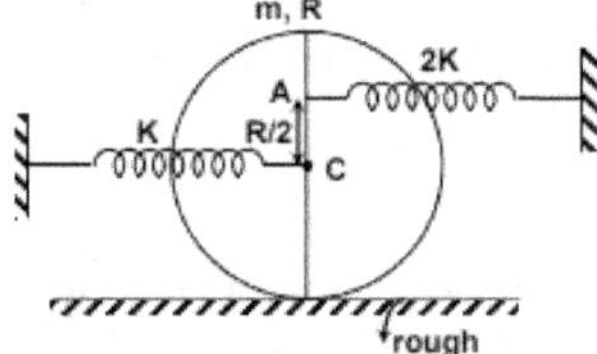

(A) $2\pi\sqrt{\dfrac{5m}{7k}}$

(B) $2\pi\sqrt{\dfrac{3m}{7k}}$

(C) $2\pi\sqrt{\dfrac{5m}{11k}}$

(D) $2\pi\sqrt{\dfrac{3m}{11k}}$

Two moles of an ideal monoatomic gas undergoes a process VT = constant. If temperature of the gas is increased by $\Delta T = 300$ K, then the ratio $\left(\dfrac{\Delta U}{\Delta Q}\right)$ is

(A) 2

(B) 3

(C) 4

(D) 6

A heat engine employing a carnot cycle with an efficiency of $\eta = 10\%$ is used as a refrigerating machine with the same thermal reservoirs. Then its coefficient of performance is

(A) 3

(B) 5

(C) 7

(D) 9

14. A string of length 50 cm and mass 2.5 g is fixed at both ends. A pipe closed at one end has a length of 85 cm. When the string vibrates in its second overtone and the air column in the pipe in its first overtone, they produce a beats frequency of 6 Hz. It is also observed that decreasing the tension in the string decreases beats frequency. Neglect the end correction in the pipe and velocity of sound in air is 340 m/s. then the tension in the string is

(A) 72 N                      (B) 52 N
(C) 26 N                      (D) 18 N

15. A car is moving with a constant velocity 10 m/s towards a stationary wall. An observer in between the car and the wall is moving with a constant velocity 2 m/s towards the stationary wall. The car blows a horn of frequency 320 Hz. The velocity of sound in air is 330 m/s. The beats frequency received by the observer is

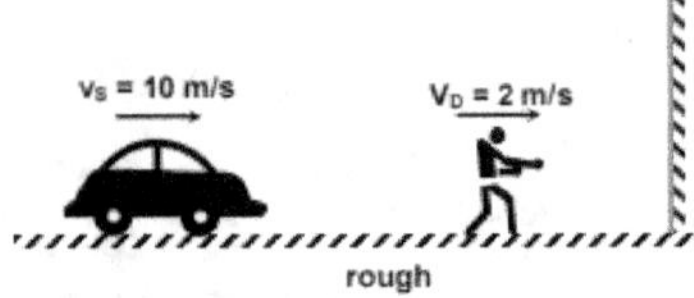


(A) 2 Hz                      (B) 4 Hz
(C) 6 Hz                      (D) 8 Hz

16. Inside a uniformly charged infinitely long cylinder of radius 'R' and volume charge density '$\rho$' there is a spherical cavity of radius 'R/2'. A point 'P' is located at a distance 2R from the axis of the cylinder as shown. Then the electric field strength at the point 'P' is

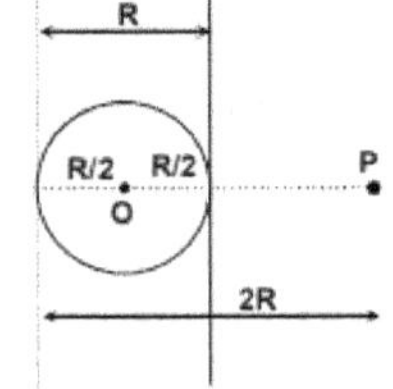


(A) $\dfrac{23\rho R}{54\varepsilon_0}$                  (B) $\dfrac{23\rho R}{108\varepsilon_0}$

(C) $\dfrac{25\rho R}{54\varepsilon_0}$                  (D) $\dfrac{25\rho R}{108\varepsilon_0}$

17. A parallel plate capacitor 'A' of capacitance 1 $\mu$F is charged to a potential drop 100 volt and then disconnected from the battery. Now the capacitor 'A' is completely filled with a dielectric slab of dielectric constant k = 4. Another parallel plate capacitor 'B' of capacitance 2 $\mu$F is charged to a potential drop 20 volt and then disconnected from the battery. Now the two capacitors 'A' and 'B' are connected with each other with opposite polarities. Then the heat dissipated after connecting the capacitors is

(A) $1.15 \times 10^{-3}$ J             (B) $1.25 \times 10^{-3}$ J
(C) $1.35 \times 10^{-3}$ J             (D) $1.45 \times 10^{-3}$ J

8. In the circuit shown, the switch 'S' is closed at t = 0. Then the current through the battery after steady state reached is
(A) 2A
(B) 4A
(C) 6A
(D) 8A

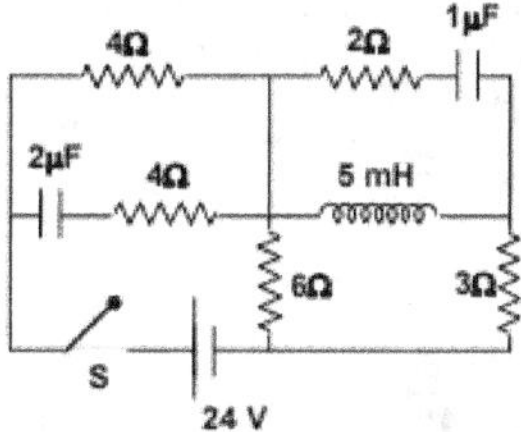

9. In the circuit shown, the switch 'S' is closed at t = 0. The capacitor is initially uncharged. Then the current through the resistor R = 8Ω at t = 0.4 × 10⁻³ sec is  (Take $e^{-2} = 0.135$)
(A) 0.18 A
(B) 0.27 A
(C) 0.54 A
(D) 0.72 A

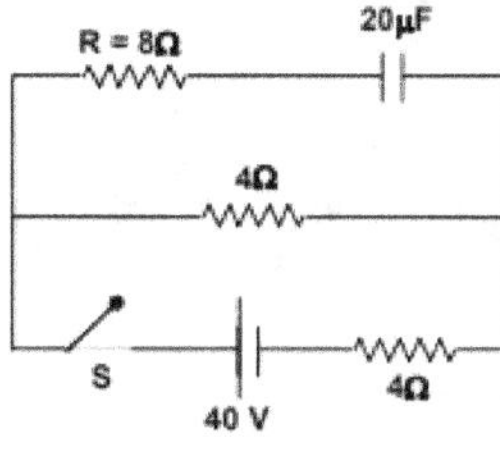

10. A uniform wire of mass '3m' and length '3l' is bent into the shape of an equilateral triangle and suspended from a horizontal axis X'X in a vertical plane as shown. A uniform vertical upward magnetic field 'B' is existing in the region. If charge 'Q' is passed almost instantaneously through the triangular loop, then the angular velocity acquired by the triangular loop immediately after charge passed is

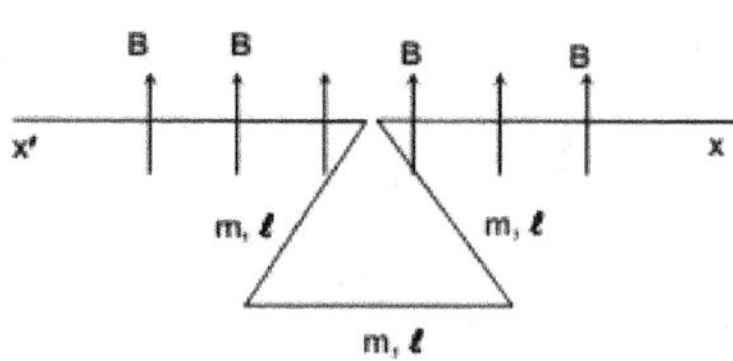

(A) $\dfrac{\sqrt{3}BQ}{5m}$

(B) $\dfrac{2\sqrt{3}BQ}{5m}$

(C) $\dfrac{3\sqrt{3}BQ}{5m}$

(D) $\dfrac{4\sqrt{3}BQ}{5m}$

21. A long wire carrying current 'I' is bent into the shape as shown in the figure. The net magnetic field intensity at the centre 'O' is

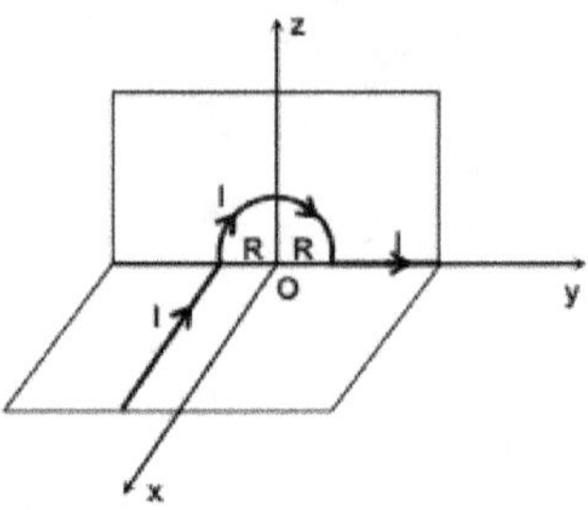

(A) $\dfrac{\mu_0 I}{4\pi R}\sqrt{1+\pi^2}$

(B) $\dfrac{\mu_0 I}{4\pi R}\sqrt{2+\pi^2}$

(C) $\dfrac{\mu_0 I}{4\pi R}\sqrt{4+\pi^2}$

(D) $\dfrac{\mu_0 I}{4\pi R}\sqrt{9+\pi^2}$

22. A time varying magnetic field $B = Kr^3 t$ is existing in a cylindrical region of radius 'R' as shown (where k is a constant). The induced electric field 'E' at $r = \dfrac{R}{2}$ is

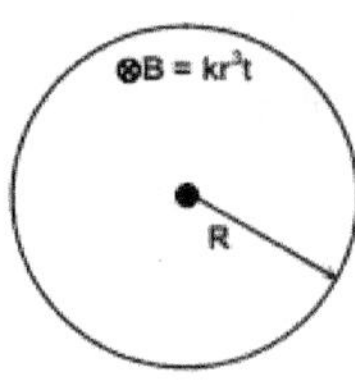


(A) $\dfrac{KR^4}{20}$

(B) $\dfrac{KR^4}{40}$

(C) $\dfrac{KR^4}{60}$

(D) $\dfrac{KR^4}{80}$

23. In the adjacent circuit, the capacitor 'A' of capacitance '3C' and the capacitor 'B' of capacitance 'C' are initially charged by potential drops $4\varepsilon$ and $8\varepsilon$ respectively and connected through an inductor of inductance 'L' with zero initial current as shown. The switch 'S' is closed at t = 0. then the maximum current through the inductor is

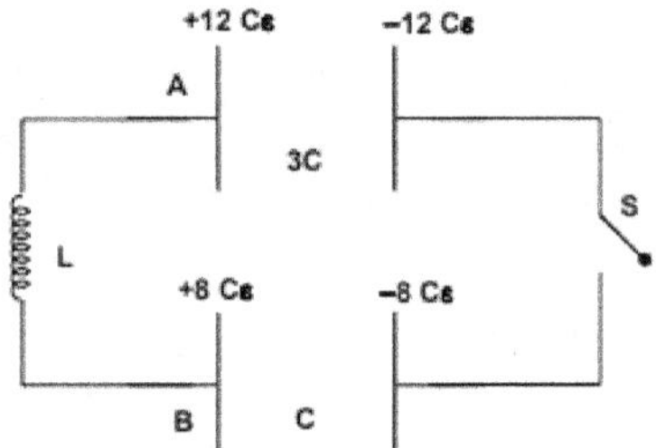


(A) $\varepsilon\sqrt{\dfrac{3C}{L}}$

(B) $\varepsilon\sqrt{\dfrac{6C}{L}}$

(C) $\varepsilon\sqrt{\dfrac{12C}{L}}$

(D) $\varepsilon\sqrt{\dfrac{15C}{L}}$

4.  Two smooth horizontal parallel conducting rails are connected with a capacitor and a resistor at the two ends as shown. A uniform conducting rod PQ of mass 'm' and length '$\ell$' is dragged with a constant horizontal force 'F'. The terminal velocity acquired by the conducting rod is

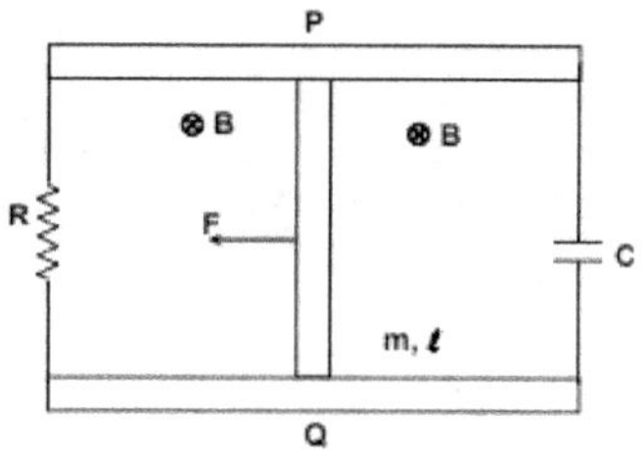


(A) $\dfrac{4FR}{B^2\ell^2}$

(B) $\dfrac{2FR}{B^2\ell^2}$

(C) $\dfrac{FR}{B^2\ell^2}$

(D) $\dfrac{FR}{2B^2\ell^2}$

5.  A plano-convex lens $(\mu = 1.5)$ of aperture diameter 8 cm has a maximum thickness of 4 mm. The focal length of the lens relative to air is

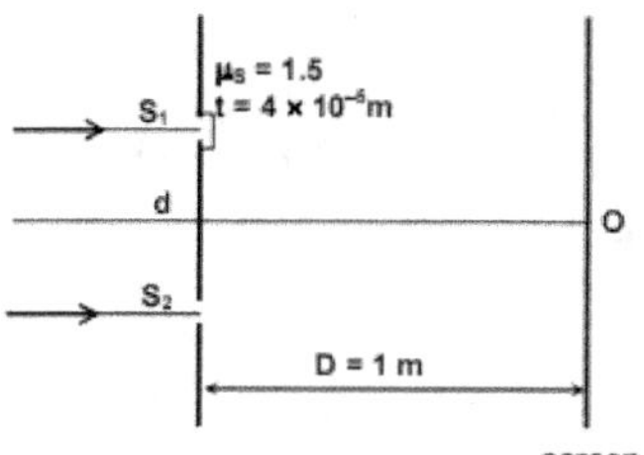

(A) + 20 cm
(B) + 40 cm
(C) + 60 cm
(D) + 80 cm

6.  A coherent parallel beam of light of wavelength 5000 Å is incident normally on the plane of slits. The slit $S_1$ is covered with a slab of refractive index $\mu_s = 1.5$ and thickness $t = 4 \times 10^{-5}$ m as shown. The separation between the slits is d = 1 mm and the separation between the plane of slits and screen is D = 1m. Then the position of the central maxima formed on the screen is

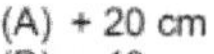



(A) 1 cm above point 'O'.
(B) 2 cm above point 'O'.
(C) 1 cm below point 'O'.
(D) 2 cm below point 'O'.

27. A thin equiconvex glass lens $(\mu_1 = 1.5)$ of radius of curvature 'R' and a thin plano-concave lens $(\mu_2 = 1.4)$ are kept in contact and the plane surface is silvered. This lens system is placed in air. When a point object is placed at a distance 20 cm from the lens system on its principal axis, the image formed by the lens system coincides with the object. Then the value of radius of curvature 'R' is

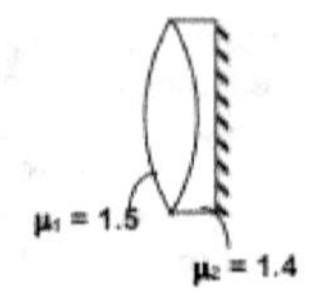


(A) 12 cm

(B) 18 cm

(C) 24 cm

(D) 30 cm

28. When the light of wavelength 400 nm is incident on a metal surface of work function 2.3 eV, photoelectrons are emitted. A fastest photoelectron combines with a $He^{2+}$ ion to form $He^+$ ion in its third excited state and a photon is emitted in this process. Then the energy of the photon emitted during combination is

(A) 4.2 eV

(B) 5.7 eV

(C) 6.5 eV

(D) 6.8 eV

29. An unpolarised light of intensity '$I_0$' is passed through the two Polaroid's placed parallel to each other and their transmission axes are inclined at an angle 45. The intensity of the polarised light after passing through the Polaroid's will be

(A) $\dfrac{I_0}{8}$

(B) $\dfrac{I_0}{4}$

(C) $\dfrac{I_0}{2}$

(D) $\dfrac{I_0}{\sqrt{2}}$

30. The pitch of a screw gauge is 1mm and its cap is divided into 100 divisions. When this screw gauge is used to measure the diameter of a wire, the main scale reading is 2mm and $58^{th}$ division of circular scale coincides with the main scale. There is no zero error in the screw gauge. Then the diameter of the wire is

(A) 2.38 mm

(B) 2.48 mm

(C) 2.58 mm

(D) 3.58 mm

# Solution

Given $v = a_0 t$

$$a_t = \frac{dv}{dt} = a_0$$

Now, $v^2 = u^2 + 2a_t S$

$$= 0 + 2a_0 \frac{\pi R}{2}$$

$$v^2 = \pi a_0 R$$

So, $a_n = \frac{v^2}{R} = \pi a_0$

The angle between the velocity vector and the acceleration vector is

$$\phi = \tan^{-1}\left(\frac{a_n}{a_t}\right) = \tan^{-1}\left(\frac{\pi a_0}{a_0}\right) = \tan^{-1}(\pi)$$

The two blocks will move with a common acceleration

$$a = \frac{100 - 30}{10} = 7 \text{ m/s}^2$$

Now, for the block 'A'

$$20 - f_s = 2 \times 7$$

$$\therefore f_s = 6N$$

Let $v$ is the velocity of the body relative to the sphere when it loses contact with the sphere.

$$mg \cos \theta - N - ma_0 \sin \theta = \frac{mv^2}{R}$$

To lose contact with the sphere, $N = 0$

$$mg \cos \theta - ma_0 \sin \theta = \frac{mv^2}{R}$$

Also, $mgR(1 - \cos \theta) + ma_0 R \sin \theta = \frac{1}{2}mv^2$

$$mg = \frac{3mv^2}{2R}$$

$$\therefore v = \sqrt{\frac{2gR}{3}}$$

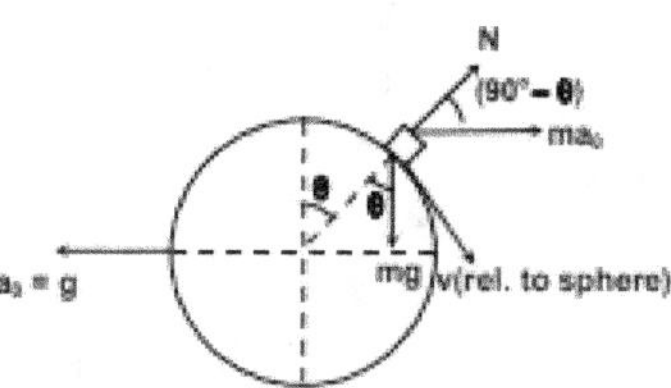


Using conservation of momentum of the system along horizontal direction

$$\frac{mv_0}{\sqrt{2}} = mv_2 - \frac{mv_1}{\sqrt{2}}$$

$$\sqrt{2}v_2 - v_1 = v_0 \qquad \ldots(i)$$

$$v_2 \cos 45° + v_1 = v_0$$

$$\frac{v_2}{\sqrt{2}} + v_1 = v_0$$

$$\sqrt{2}v_2 + 2v_1 = 2v_0 \qquad \ldots(ii)$$

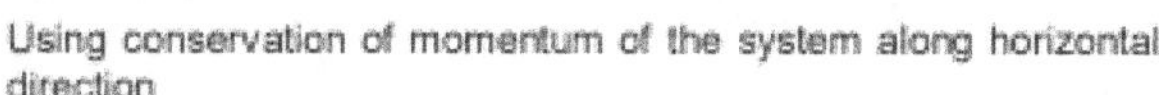


Solving (i) and (ii), we get

$3v_1 = v_0$

$v_1 = \dfrac{v_0}{3} = \dfrac{6}{3} = 2 \ m/s$

5.   $F - f_s = ma$     ...(i)

$\dfrac{FR}{3} + f_s R = \dfrac{2}{3}mR^2\alpha$

$\dfrac{F}{3} + f_s = \dfrac{2}{3}ma$    [since, $a = R\alpha$]    ...(ii)

Solving (i) and (ii), we get

$\dfrac{4F}{3} = \dfrac{5}{3}ma$

$a = \dfrac{4F}{5m}$

Now, from equation (i)

$f_s = F - ma = F - \dfrac{4F}{5} = \dfrac{F}{5}$

Hence, $f_s \le \mu N$

$F/5 \le \mu mg$

$F \le 5\,\mu mg$

$\therefore F_{max} = 5 \times 0.4 \times 20 = 40 \ N$

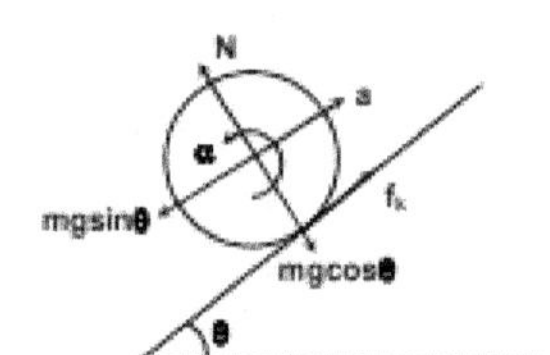


6.   For toppling of the block

$ma\dfrac{h}{2} > mg\dfrac{w}{2}$

$a > \dfrac{w}{h}g$

$a > \dfrac{15}{60} \times 10$

$a > 2.5 \ m/s^2$



7.   $N = mg\cos\theta$

$f_k = \mu N = 2\tan\theta\, mg\cos\theta = 2mg\sin\theta$

$f_k - mg\sin\theta = ma$

$2mg\sin\theta - mg\sin\theta = ma$

$\therefore \quad a = g\sin\theta$

Now, $\tau_{cm} = I_{cm}\alpha$

$f_k R = \dfrac{2}{5}mR^2\alpha$

$2mg\sin\theta = \dfrac{2}{5}mR\alpha$

$\alpha = \dfrac{5g\sin\theta}{R}$

When the rolling without slipping starts

$v = \omega R$

$v_0 + at = (\omega_0 - \alpha t)R$

$v_0 + g\sin\theta t = \omega_0 R - 5g\sin\theta t$

$v_0 + 6g\sin\theta t = 2v_0$

$\therefore \quad t = \dfrac{v_0}{6g\sin\theta}$

Force of surface tension on each wire is $F = 2T\ell$

$\dfrac{md^2x}{dt^2} = F - 2Kx$

$= -2K\left(x - \dfrac{F}{2k}\right)$

$\dfrac{d^2x}{dt^2} = -\dfrac{2k}{m}\left(x - \dfrac{F}{2k}\right)$

$\therefore$ Amplitude of SHM,

$A = \dfrac{F}{2k} = \dfrac{2T\ell}{2k} = \dfrac{T\ell}{k}$

$\therefore$ Maximum elongation produced in the spring is

$\Delta\ell_{max} = 4A = \dfrac{4T\ell}{k}$

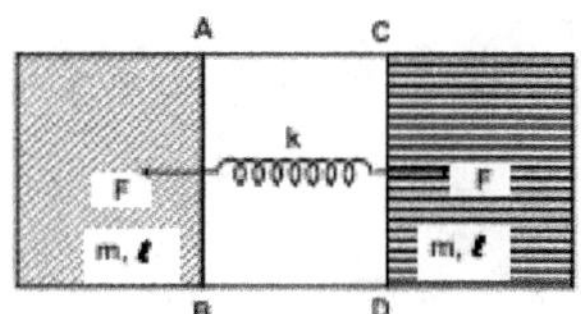

$2F\sin\theta = mg$

$F = \dfrac{mg}{2\sin\theta} = \dfrac{mg\ell}{2x} \quad [\because \theta \text{ is small}]$

$\Delta\ell = \left(\sqrt{\ell^2 + x^2} - \ell\right)$

$= \ell\left[\left(1 + \dfrac{x^2}{\ell^2}\right)^{1/2} - 1\right]$

$\Delta\ell = \ell\left[1 + \dfrac{1}{2}\dfrac{x^2}{\ell^2} - 1\right] \quad [\because x \ll \ell]$

$\therefore \quad \dfrac{\Delta\ell}{\ell} = \dfrac{x^2}{2\ell^2}$

Now $\dfrac{F}{A} = Y\dfrac{\Delta\ell}{\ell}$

$F = AY\dfrac{\Delta\ell}{\ell}$

$\dfrac{mg\ell}{2x} = AY\left(\dfrac{x^2}{2\ell^2}\right)$

$x^3 = \dfrac{mg\ell^3}{AY}$

$\therefore \quad x = \ell\left(\dfrac{mg}{AY}\right)^{1/3}$

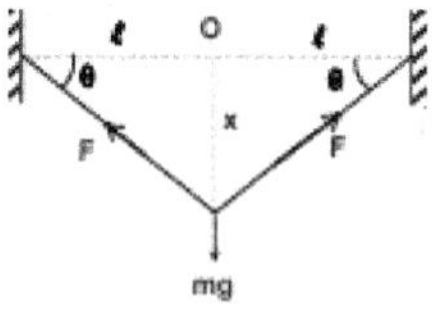

0. Using conservation of energy

$\dfrac{1}{2}mv^2 = -\dfrac{GMm}{R} - \left(\dfrac{-11GMm}{8R}\right)$

$$\frac{1}{2}mv^2 = \frac{3GMm}{8R}$$

$$\therefore \ v = \sqrt{\frac{3GM}{4R}}$$

11.  $$\frac{3mR^2}{2}\frac{d^2\theta}{dt^2} = -kR^2\theta - 3kR\theta\left(\frac{3R}{2}\right)$$

$$\frac{3mR^2}{2}\frac{d^2\theta}{dt^2} = -\frac{11kR^2\theta}{2}$$

$$\frac{d^2\theta}{dt^2} = -\left(\frac{11K}{3m}\right)\theta$$

$$\therefore \ \text{Time Period}, \quad T = 2\pi\sqrt{\frac{3m}{11k}}$$

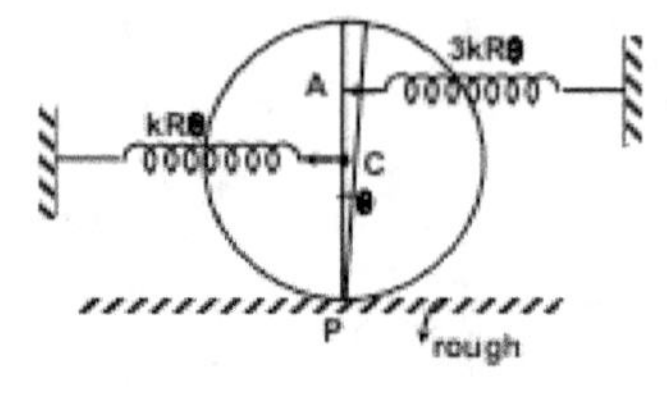


12.  Given, $VT = $ constant

$PV^2 = $ constant

Polytropic constant, $x = 2$

$\therefore$ Molar heat capacity,

$$C = C_v + \frac{R}{1-x}$$

$$= \frac{3R}{2} + \frac{R}{1-2}$$

$$= \frac{R}{2}$$

$$\therefore \ \Delta Q = nC\Delta T = 2 \times \frac{R}{2} \times 300 = 300R$$

$$\Delta U = nC_v\Delta T = 2 \times \frac{3R}{2} \times 300 = 900R$$

$$\therefore \ \frac{\Delta U}{\Delta Q} = \frac{900R}{300R} = 3$$

13.  Efficiency of carnot engine is, $\eta = 1 - \dfrac{T_2}{T_1}$

$$0.1 = 1 - \frac{T_2}{T_1}$$

$$\therefore \ \frac{T_2}{T_1} = 0.9$$

Coefficient of performance (COP) of a refrigerator $= \dfrac{T_2}{T_1 - T_2} = \dfrac{0.9\,T_1}{T_1 - 0.9\,T_1} = 9$

$$\therefore \ \text{COP} = 9$$

14.  The frequency of vibration of air column in the pipe is $f_1 = \dfrac{3v}{4\ell} = \dfrac{3 \times 340}{4 \times 0.85} = 300$ Hz.

The frequency of vibration of string is $f_2 = f_1 + 6 = 300 + 6 = 306$ Hz

Now, $f_2 = \dfrac{3}{2\ell}\sqrt{\dfrac{F}{\mu}}$

$$306 = \frac{3}{2 \times 0.5}\sqrt{\frac{F \times 0.5}{2.5 \times 10^{-3}}}$$

$$306 = 30\sqrt{2F}$$

$$\sqrt{2F} = 10.2 \quad \therefore \quad F = 52 \text{ N}$$

5. $f_1 = \left(\dfrac{v - v_0}{v - v_s}\right) f$ and the frequency of sound received by the observer after reflection from the

   stationary wall is $f_2 = \left(\dfrac{v + v_0}{v - v_s}\right) f$

   $\therefore$ Beats frequency received by the observer, $f_b = f_2 - f_1 = \left(\dfrac{v + v_0}{v - v_s}\right) f - \left(\dfrac{v - v_0}{v - v_s}\right) f$

   $\therefore \quad f_b = \dfrac{(2v_0)f}{(v - v_s)} = \dfrac{2 \times 2}{(330 - 10)} \times 320 = 4\,\text{Hz}.$

6. The electric field strength at point 'P' is

   $E = E_1 - E_2$

   $= \dfrac{\rho R}{4\varepsilon_0} - \dfrac{\rho R}{54\varepsilon_0}$

   $E = \dfrac{25\rho R}{108\varepsilon_0}$

7. $Q_A = 1 \times 100 = 100\,\mu C$

   $Q_B = 2 \times 20 = 40\,\mu C$

   When the two capacitors are connected with opposite polarities.

   $q_A = 40\,\mu C, \; q_B = 20\,\mu C$

   The heat dissipated after the two capacitors connected

   $\Delta H = U_i - U_f = \left[\dfrac{(100)^2}{2 \times 4} + \dfrac{(40)^2}{2 \times 2}\right] - \dfrac{(60)^2}{2 \times 6}$

   $= (1250 + 400) - 300 = 1350\,\mu J = 1.35 \times 10^{-3}\,\text{J}$

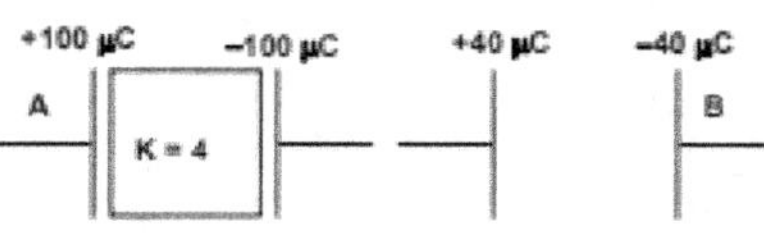


8. In steady state, the equivalent circuit is shown.

   $R_{eq} = 4 + \left(\dfrac{6 \times 3}{6 + 3}\right)$

   $R_{eq} = 4 + 2 = 6\,\Omega$

   $\therefore \quad I_s = \dfrac{\varepsilon}{R_{eq}} = \dfrac{24}{6} = 4\,\text{A}.$

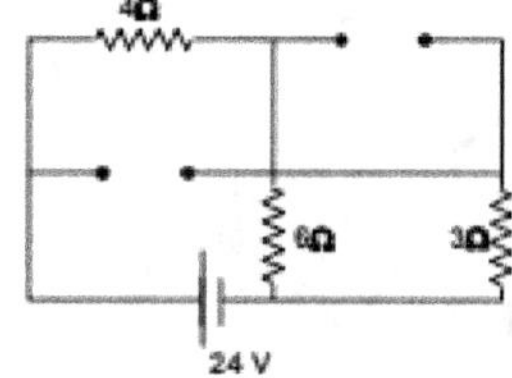

19.     Time constant, $\tau = R_{eq}C_{eq}$

$$= 10 \times 20 \times 10^{-6}$$
$$= 0.2 \times 10^{-3} \text{ sec}$$

The current through the battery just after closing the switch is

$$I_0 = \frac{40 \times 3}{20} = 6A$$

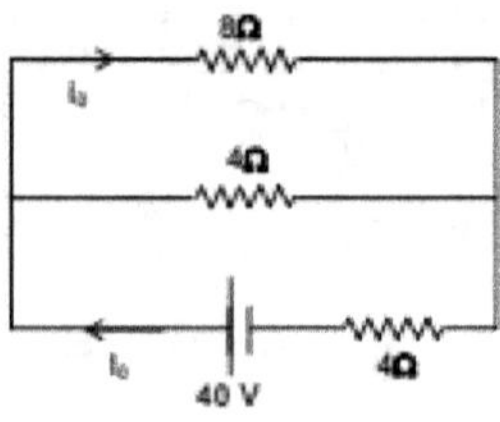


The current through R = 8 Ω register at t = 0 is

$$i_0 = \frac{1}{3} \times I_0 = \frac{1}{3} \times 6 = 2A$$

The current through R = 8Ω resistor at time 't'

$$i = i_0 e^{-t/\tau}$$

$$i = 2\, e^{-t/0.2 \times 10^{-3}}$$

$\therefore$ at $t = 0.4 \times 10^{-3}$ sec.

$$i = 2e^{-2} = 2 \times 0.135 = 0.27\,A$$

20.     $\tau = \dfrac{\sqrt{3}}{4}Bi\ell^2$

Now, $\int \tau dt = I\omega$

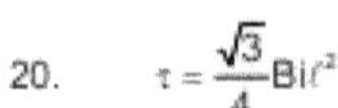

$$\frac{\sqrt{3}}{4}B\ell^2 \int i\,dt = \frac{5m\ell^2\omega}{4}$$

$$\frac{\sqrt{3}}{4}B\ell^2 Q = \frac{5m\ell^2}{4}\omega$$

$\therefore$ $\omega = \dfrac{\sqrt{3}BQ}{5m}$

21.     The net magnetic field intensity at the centre 'O' is

$$\vec{B}_0 = \frac{\mu_0 I}{4\pi R}\left(-\hat{k}\right) + \frac{\mu_0 I}{4R}\left(-\hat{i}\right)$$

$\therefore$ $B_0 = \dfrac{\mu_0 I}{4\pi R}\sqrt{1 + \pi^2}$

22.     Using Faraday's law

$$E2\pi r = -\frac{d}{dt}\left[-\int_0^r B2\pi r\,dr\right]$$

$$= \frac{d}{dt}\left[\int_0^r Kr^3 t2\pi r\,dr\right]$$



$$E2\pi r = \frac{d}{dt}\left(2\pi kt\frac{r^5}{5}\right)$$

$E = \dfrac{Kr^4}{5}$, when $r \le R$

$\therefore$ at $r = \dfrac{R}{2}$,    $E = \dfrac{KR^4}{80}$

When the current through the inductor is maximum,
$$V_L = L\frac{di}{dt} = 0$$

Now, using energy conservation

$$\frac{1}{2}Li_{max}^2 = \left(\frac{1}{2}\times 3C \times 16\varepsilon^2 + \frac{1}{2}C\times 64\varepsilon^2\right) - \frac{1}{2}(3C+C)25\varepsilon^2$$

$$\frac{1}{2}Li_{max}^2 = 56\,C\varepsilon^2 - 50C\varepsilon^2$$

$$\frac{1}{2}Li_{max}^2 = 6C\varepsilon^2$$

$$\therefore \quad i_{max} = \varepsilon\sqrt{\frac{12C}{L}}$$

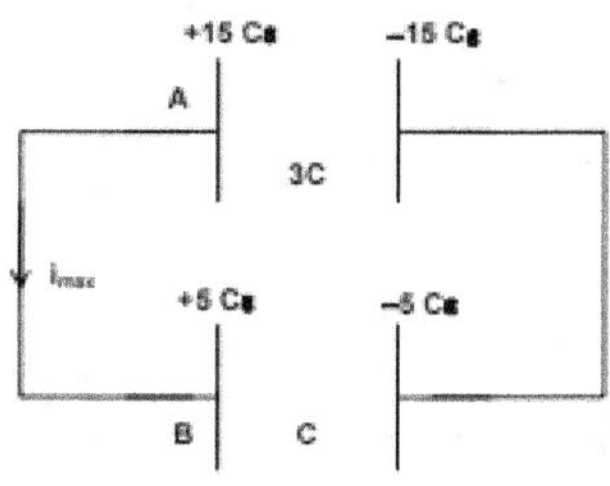


When the terminal velocity is acquired

$$i = \frac{Bv\ell}{R}$$

$$F = Bi\ell$$

$$F = \frac{B^2\ell^2 v}{R}$$

$$\therefore \quad v = \frac{FR}{B^2\ell^2}$$

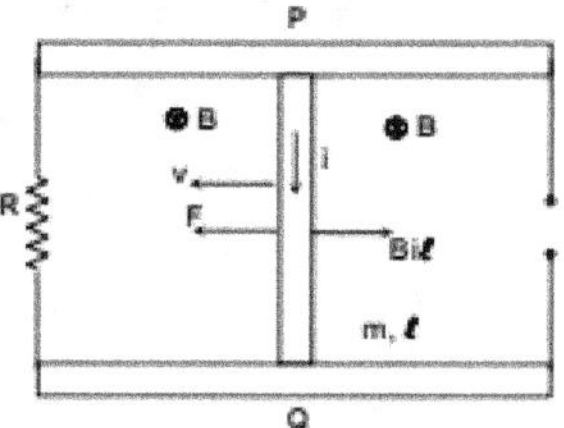


$$R^2 = (R-t)^2 + r^2$$

$$R^2 = R^2 + t^2 - 2Rt + r^2$$

$$r^2 = 2Rt \quad (t^2 \text{ is neglected})$$

$$R = \frac{r^2}{2t} = \frac{4\times 4}{2\times 0.4} = 20 \text{ cm}$$

$$R = 20 \text{ cm}$$

$$\therefore \quad \frac{1}{f} = (1.5-1)\left(\frac{1}{\infty} - \frac{1}{-20}\right)$$

$$\frac{1}{f} = \frac{0.5}{20} \quad \therefore f = +40 \text{ cm}$$

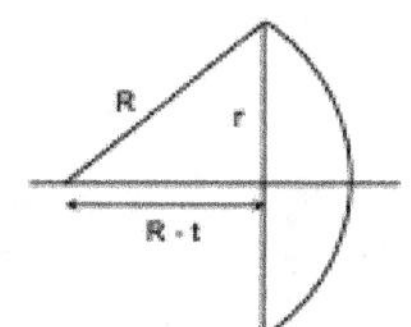


For central maxima $d\sin\theta = (\mu_a - 1)t$

$$1\times 10^{-3}\sin\theta = (1.5-1)4\times 10^{-5}$$

$$1\times 10^{-3}\sin\theta = 2\times 10^{-5}$$

$$\sin\theta = \frac{1}{50}$$

$$\tan\theta = \frac{1}{50} \quad (\text{for small } '\theta')$$

$$\frac{y}{D} = \frac{1}{50}$$

$$\therefore \quad y = \frac{D}{50} = \frac{100}{50} \text{ cm} = 2 \text{ cm (above point 'O')}$$

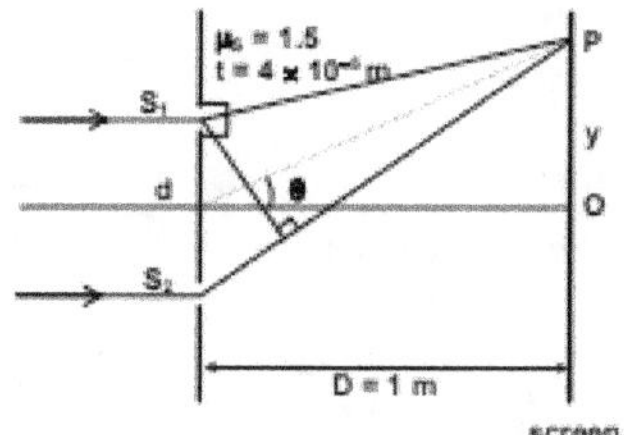

27.  $\dfrac{1}{f_1} = (1.5 - 1)\left(\dfrac{1}{R} + \dfrac{1}{R}\right) = \dfrac{1}{R}$

$\therefore \quad f_1 = R$

$\dfrac{1}{f_2} = (1.4 - 1)\left(\dfrac{1}{-R} - \dfrac{1}{\infty}\right)$

$\dfrac{1}{f_2} = -\dfrac{0.4}{R}$

$-\dfrac{1}{F} = \dfrac{2}{f_1} + \dfrac{2}{f_2} - \dfrac{1}{f_m}$

Since image coincides with the object,
$F = -10$ cm

$\dfrac{-1}{-10} = \dfrac{2}{R} + 2\left(\dfrac{-0.4}{R}\right) - \dfrac{1}{\infty}$

$\dfrac{1}{10} = \dfrac{2}{R} - \dfrac{0.8}{R}$

$\dfrac{1}{10} = \dfrac{1.2}{R}$

$\therefore \quad R = 12$ cm

28.  Energy of each photon

$E = \dfrac{hC}{\lambda} = \dfrac{1240}{400} = 3.1\,\mathrm{eV}$

Now, $K_{max} = E - \phi$

$= 3.1 - 2.3$

$\therefore \quad K_{max} = 0.8\,\mathrm{eV}$

Energy of $He^+$-ion in third excited state

$E_4 = -13.6 \times \dfrac{4}{(4)^2} = -3.4\,\mathrm{eV}$

Now, the energy of the photon emitted during combination,

$\Delta E = K_{max} - E_4$

$= 0.8 - (-3.4)$

$= 4.2\,\mathrm{eV}$

29.  The intensity of the light after passing through the first Polarised $= \dfrac{I_0}{2}$. Now, the intensity of the light after getting passed through the second Polaroid will be

$I = \dfrac{I_0}{2}\cos^2\theta$

$= \dfrac{I_0}{2}\cos^2 45^\circ = \dfrac{I_0}{4}$

30.  The pitch of the screw gauge, $P = 1$ mm

Least count, $L.C. = \dfrac{P}{N} = \dfrac{1\,mm}{100} = 0.01\,mm$

$\therefore \quad$ Diameter of the wire, $d = 2mm + 58 \times L.C.$

$= 2mm + 58 \times 0.01\,mm$

$= 2.58\,mm$